Organic reactions and orbital symmetry

Cambridge Chemistry Texts

GENERAL EDITORS

E. A. V. Ebsworth, Ph.D.
Professor of Inorganic Chemistry,
University of Edinburgh

D. T. Elmore, Ph.D.
Reader in Biochemistry,
The Queen's University of Belfast

P. J. Padley, Ph.D.
Lecturer in Physical Chemistry,
University College of Swansea

K. Schofield, D.Sc.
Reader in Organic Chemistry,
University of Exeter

Organic reactions and orbital symmetry

T. L. GILCHRIST
AND
R. C. STORR

*Department of Organic Chemistry,
The Robert Robinson Laboratories,
University of Liverpool*

CAMBRIDGE
at the University Press 1972

Published by the Syndics of the Cambridge University Press
Bentley House, 200 Euston Road, London NW1 2DB
American Branch: 32 East 57th Street, New York, N.Y.10022

© Cambridge University Press 1972

Library of Congress Catalogue Card Number: 71-161286

ISBN: 0 521 08249 8 Clothbound
        0 521 09658 8 Paperback

Printed in Great Britain
by William Clowes & Sons Limited
London, Colchester and Beccles

Contents

Preface

The application of the concept of orbital symmetry to organic chemistry by R. B. Woodward and R. Hoffmann has proved to be a major theoretical advance, in that it has succeeded in bringing together and rationalising diverse areas of the subject. In particular it has provided the basis for a unified mechanistic approach to cycloadditions and molecular rearrangements; partly as a result of the stimulus of the new theory, the importance of such reactions is now rightly recognised. For historical reasons these reactions have not usually been treated as fully as their importance warrants in student texts. Now that the concept of orbital symmetry control is well established it seems appropriate to present an account of these reactions within a modern mechanistic framework.

The major part of the book (chapters 3 to 7) is devoted to a descriptive account of rearrangements and cycloadditions. The aim has been to illustrate the scope and synthetic utility of the reactions as well as to discuss their mechanisms. Chapter 1 gives an introduction to the types of mechanisms by which such reactions can occur, and to the experimental methods available for establishing the mechanisms. In particular, the important distinction between stepwise and concerted processes is emphasised. The treatment in chapter 1 is elementary and descriptive, the aim being to provide a brief revision of certain terms and concepts which are used throughout the rest of the book. Chapter 2 is a comparative account of the various approaches to the theory of concerted reactions. In a brief final chapter (chapter 8) we have speculated on possible extensions of the theory to other types of concerted processes. Throughout the book, thermally induced reactions are given more detailed treatment than photochemical and catalysed reactions, for which the applications of the theory are, as yet, less firmly established. References are given at the end of each chapter to relevant reviews, and also to important original papers and to work which has appeared since the publication of the reviews.

We are indebted to our colleagues Dr D. Bethell and Dr M. J. P. Harger, and to the Series Editor, Dr K. Schofield, for their constructive criticisms. We are also grateful to Professor C. W. Rees, who not only made helpful criticisms of the text, but also was a source of advice and encouragement during its preparation.

T. L. GILCHRIST

R. C. STORR

August 1971

1 Classification and investigation of reaction mechanisms

The investigation of reaction mechanisms by organic chemists during the last fifty years or so has provided a satisfying and coherent picture of organic reactions. As ideas of mechanism developed, two broad classes of reactions, involving either ionic species or free radicals, were recognised. A series of practical tests could be applied to investigate mechanisms; for example, the effect of changing solvent or substituents on the reaction rate and the trapping of reaction intermediates. Many reactions could confidently be classified on this basis as ionic or radical processes. A third group fell outside this classification in that they seemed to be insensitive to the usual tests of mechanism; they appeared not to involve intermediates but proceeded by a reorganisation of electrons through a four- or six-centre cyclic transition state, and were therefore classed as *concerted* (one-step) processes.

Since 1965, Woodward and Hoffmann have developed a general theory of concerted reactions, by applying the principles of orbital symmetry.[1] The theory enables us to predict when a concerted reaction is likely to be energetically feasible. This has provided the impetus for a reinvestigation and reinterpretation of many reactions, especially the group which apparently involved cyclic transition states. This new theory emphasises the distinction between stepwise and concerted processes, and so it is essential to look at the criteria which can be used to distinguish them. The purpose of this chapter is to describe these criteria. Before this is done, a brief outline of the energetics of chemical reactions will be given.

1.1. Reaction rates and equilibria.[2,4] Two factors are important in determining the thermodynamic stability of a molecule: the stabilising energy resulting from the formation of bonds between the atoms and the destabilising energy due to the loss of freedom involved in constraining

1

the atoms within the molecular structure. The thermodynamic function which embraces both of these factors is the *free energy* (*G*) now called Gibbs Function in SI terminology. Free energy is the fundamental quantity which controls the feasibility and the rate of all reactions.

The reactants in a chemical system usually have a higher free energy than the products and the forward reaction can then, in principle, proceed spontaneously. If the products have a higher free energy than the reactants, energy has to be supplied from some external source for the forward reaction to go. However, even when the reactants have a higher free energy than the products, the forward reaction is rarely spontaneous, but proceeds at a finite rate which may be extremely slow except at high temperatures. Theories which attempt to explain this use the concept of an *energy barrier* which has to be surmounted.

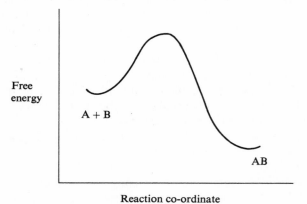

Fig. 1.1

A reaction involves a gradual breaking of some bonds and/or formation of others. According to *transition state theory*, the reaction course can be considered as an infinite series of equilibria between one structure and another. The structure of highest free energy on the reaction path is called the *transition state* or *activated complex*. Although there are an infinite number of such routes from reactants to products, we are concerned only with the one of lowest energy – the so-called *reaction profile*. The free energy in such a process is illustrated in fig. 1.1 for the simple reaction

$$A + B \rightleftharpoons AB$$

The free energy is shown as a function of *reaction co-ordinate*. This is a rather loose term which indicates progress of change from reactants to products: for the example shown, it could be the length of the A—B bond.

The equilibrium constant K for any reversible reaction is determined by the difference in free energy ΔG between reactants and products. For the system

$$A + B \rightleftharpoons AB$$

$$K = \frac{[AB]}{[A][B]}$$

and $$\Delta G = -RT \ln K = -RT \ln \frac{[AB]}{[A][B]}$$

The relative proportions of products and reactants at equilibrium are therefore determined by their difference in free energy.

Transition state theory also enables the rate of a reaction to be linked to free energy differences. The theory assumes an equilibrium between the reactants $A + B$ and the activated complex $A \ldots B$. The concentration of the activated complex is therefore determined by ΔG^{+}, the difference in free energy between the reactants and the transition state. ΔG^{+} is called the *free energy of activation*, because it is the free energy barrier which must be surmounted for reaction to occur. In transition state theory the further assumption is made that all activated complexes break down at the same rate. This can be shown to be reasonable by a statistical argument which gives kT/h as the rate constant for the break-down (k = Boltzmann's constant; h = Planck's constant). Thus, the rate of the forward reaction $A + B \rightarrow AB$ is

$$\frac{akT}{h} [A\ldots B]$$

where a is the fraction of activated complex passing on to product; it is assumed to be close to 1. Putting

$$K^{+} = \frac{[A\ldots B]}{[A][B]}$$

this expression becomes

$$\text{rate} = \frac{kT}{h} \cdot K^{+}[A][B]$$

The rate constant for the reaction k_r is therefore

$$k_r = \frac{kT}{h} \cdot K^{+} = \frac{kT}{h} \cdot e^{-\Delta G^{+}/RT}$$

The rate constant at any particular temperature is thus determined by the free energy of activation ΔG^{+}. The rate constant for the reverse reaction is similarly determined by the difference in free energy between products and transition state.

The discussion so far has concerned *reversible* reactions, but in fact not all reactions are found experimentally to be reversible. An *irreversible* reaction is one in which ΔG^{+} for the reverse reaction is large compared with ΔG^{+} for the forward reaction so that the rate of the reverse reaction is negligible in comparison with the rate of the forward reaction.

It is useful to split ΔG^{+} into its component enthalpy (ΔH^{+}) and entropy (ΔS^{+}) factors:

$$\Delta G^{+} = \Delta H^{+} - T\Delta S^{+}$$

Enthalpy is essentially a bond energy term and entropy reflects the ordering of the system, so that ΔH^{+} and ΔS^{+} give more insight into the nature of the transition state. This idea of the energy barrier for a reaction as a combination of two factors is derived naturally from an alternative, and pictorially simpler, approach to the theory of reaction rates – *collision theory*. This assumes that only that fraction of molecules which collide with sufficient energy to surmount the transition state barrier can react (the ΔH term), and that even sufficiently energetic collisions will only lead to reaction if the colliding molecules are correctly aligned (the ΔS term).

Experimental values of ΔH^{+}, ΔS^{+} and ΔG^{+} are obtained by measuring the reaction rate constant at several different temperatures. The *Arrhenius equation*

$$k_{r} = Ae^{-E_a/RT} \qquad (k_{r} = \text{rate constant})$$

is an empirical relationship between rate constant and temperature which fits well for most reactions. A and E_a are constants which are, to a first approximation, independent of temperature. E_a is called the *Arrhenius activation energy* and A the *pre-exponential factor*. It can be shown that for a reaction in solution

$$\Delta H^{+} = E_a - RT$$

and for a gas phase reaction

$$\Delta H^{+} = E_a - nRT \qquad (n = \text{molecularity}).$$

At room temperature RT is about 0.6 kcal mol^{-1} (2.5 kJmol^{-1}) so there is little difference in the value of E_a and ΔH^{+}.

Since $\Delta G^{+} = \Delta H^{+} - T\Delta S^{+}$, the expression for the rate constant for a reaction derived from transition state theory

$$k_{\mathrm{r}} = \frac{kT}{h} \cdot \mathrm{e}^{-\Delta G^{+}/RT}$$

can be rewritten in a form very similar to that of the empirical Arrhenius equation:

$$k_{\mathrm{r}} = \left(\frac{kTe}{h} \cdot \mathrm{e}^{\Delta S^{+}/R} \right) \mathrm{e}^{-E_{a}/RT}$$

the expression within the brackets being equivalent to the pre-exponential factor A. Thus, ΔH^{+} and ΔS^{+} can be obtained from the experimentally determined values of E_{a} and A, respectively.

1.2. Concerted and stepwise reactions. Reactions in which more than one bond is broken or formed can be divided into two classes. The first is one in which all the bond forming and breaking processes occur simultaneously so that a one-step transformation of reactants to products occurs without the intervention of an intermediate. Such reactions

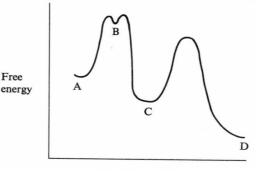

Reaction co-ordinate

Fig. 1.2

are called *concerted* or *multicentre* since the bond changes occur in concert or at the same time at more than one centre. The energy curve for such processes is shown in fig. 1.1; it involves only one energy barrier and one transition state.

The second broad class of reactions is one in which the bond forming and breaking processes occur consecutively so that one or more intermediates is involved. These intermediates may be stable molecules capable of isolation, or they may be highly reactive species of only transient existence. Where the intermediate is a stable molecule it is often more convenient to consider the overall process as two or more consecutive concerted reactions. Where the intermediates are unstable the process is normally considered as one reaction which proceeds in a stepwise manner. This distinction is purely arbitrary and one of convenience.

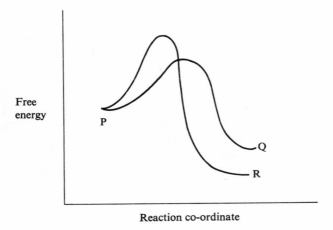

Fig. 1.3

The energy curve for a simple stepwise reaction could be as in fig. 1.2. In such a reaction C is formed from A via an unstable intermediate B which has a discrete, though short, lifetime. The instability of B is indicated by the fact that it lies in a shallow energy well. It has only a small energy barrier to surmount in order to pass over to products or to revert to reactants. The more stable such an intermediate is, the deeper the energy well. C is relatively stable and lies in a deep energy well.

Although C is the first isolable product to be formed, it may subsequently pass over to the more stable D in the reaction conditions. C is called the *kinetically controlled* product of reaction of A. D, which is the product isolated after the system reaches equilibrium, is called the *thermodynamically controlled* product. This is the type of situation where kinetic and thermodynamic control is most frequently encountered.

Kinetic and thermodynamic control can also operate in systems of the type $Q \rightleftharpoons P \rightleftharpoons R$. If the reactions $P \rightarrow Q$ and $P \rightarrow R$ have energy profiles as shown in fig. 1.3, Q is formed faster than R because the energy barrier for its formation is lower than that for R. However, if the reacting system reaches thermodynamic equilibrium, the proportions of products Q and R will be determined by their relative free energies, and R will predominate. In this type of situation Q is the kinetically controlled product and R is the thermodynamically controlled product.

If a reaction is reversible and a particular pathway is energetically the most favourable route from reactants to products, the lowest energy pathway for the reverse reaction will be along the same route but with all bond making and breaking processes reversed. Forward and reverse reactions will therefore pass through the same transition states and will involve the same mechanism. This is the *principle of microscopic reversibility*.

Throughout the book we are concerned with reactions in which cyclic structures are formed or cleaved or which, at least formally, involve a cyclic transition state. The extreme mechanisms by which such reactions can occur can conveniently be discussed with reference to a hypothetical fragmentation:

$$\begin{matrix} a\!-\!b \\ | \quad | \\ c\!-\!d \end{matrix} \longrightarrow \begin{matrix} a\!=\!b \\ c\!=\!d \end{matrix}$$

Concerted mechanism. In the concerted fragmentation the breaking of the a—c bond is coupled with, and depends on, the breaking of the b—d bond, so that both processes are going on at the same time. The overall reaction does not involve an intermediate and the energy profile is of the type shown in fig. 1.1. There is only one energy barrier with a transition state in which both σ bonds are partially broken and the new π bonds in a=b and c=d are partially formed.

The two σ bonds can break simultaneously and at exactly the same rate. It is quite reasonable to expect that the two bonds do not break at the same rate, however, especially if the reactant is unsymmetrical. In the latter case the two bonds will be broken to different extents in the transition state. However, no matter how lop-sided the transition state may be, the process is still considered to be concerted if the breaking of the a—c bond is coupled with and controlled by the breaking of the b—d bond (fig. 1.4).

In an unsymmetrical concerted process the transition state has some of the character of the intermediates which are involved in truly stepwise processes: the unequal rates of formation of the two bonds leads to the development of dipolar or diradical character in the transition state.

Woodward and Hoffmann[1] have introduced the term *pericyclic* to cover all concerted reactions which involve a cyclic transition state, and define it as follows: 'a pericyclic reaction is a reaction in which all first order changes in bonding relationships take place in concert on a closed curve'.

Fig. 1.4

Stepwise mechanism. The bonds a—c and b—d may break in two successive independent steps. The reaction then involves an intermediate in which only one of the σ bonds is broken. The energy profile will be as in fig. 1.5, with two transition states.

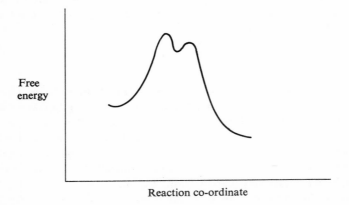

Free
energy

Reaction co-ordinate

Fig. 1.5

Stepwise breaking and formation of bonds can be detected experimentally (§ 1.3). The intermediates fall into two categories; those which are highly polar, and those which are essentially non-polar. The first type is normally considered to be a *zwitterion*; that is, a species bearing both a positive and a negative charge; and the second type, a *diradical*.

Zwitterions. In the fragmentation shown a zwitterion results from heterolytic cleavage of one of the bonds. Such a process is favoured if the resulting charges are stabilised inductively or mesomerically. Collapse of the zwitterion to re-form the bond b—d gives back the starting material.

Other courses are open to the zwitterion; the bond a—c may cleave, or the zwitterion may react with other species present. Bond rotation about the a—b, a—c, or c—d bonds in the intermediate may occur before re-closure or cleavage. It is the experimental observation of these competing processes which enables the reaction to be identified as a stepwise one.

Fig. 1.6

Diradicals. If the bond b—d cleaves homolytically, this leads to a diradical intermediate. Such a process is favoured if the structure is such that the separate monoradicals are stabilised by delocalisation. As the separation of the centres b and d increases, the correlation between the electrons decreases. At some point the separation is sufficiently large for the species to be considered as a diradical rather than as a vibrationally excited ring. Conversely, a diradical · b—a—c—d · only becomes equivalent to a vibrationally excited ring when conformational changes bring b and d within bonding distance. The diradical is therefore a discrete intermediate; it can re-close, or fragment by cleavage of the bond a—c, both processes competing with bond rotation.

If the centres bearing the unpaired electrons in a diradical are well separated (for example, by a long carbon chain) there is little interaction between them and each might be expected to act as an independent monoradical function. Usually the centres are close enough for intra-molecular interaction to preclude their reaction as separate chemical entities.

The concept of electron spin has proved very useful to organic chemists, and the question of relative spins of the unpaired electrons in a diradical is often alluded to. This is reasonable if the separation between the radical centres is small, but not otherwise. The species is referred to as a singlet if the electron spins are opposed, but as a triplet if the spins are parallel. As the bond b—d is cleaved, the electrons initially have antiparallel spins, so that the radical should be generated as a singlet. This singlet state may not be the ground state of the intermediate, however. The relationship between structure and reactivity of diradicals is a complex one, and further discussion is deferred until § 6.3.

Photochemical reactions.[3] In a photochemical reaction, a reactant absorbs radiation (generally ultraviolet) which causes excitation of an electron to a higher vacant orbital. Most frequently, lone pair electrons or π electrons are promoted to low lying π^* antibonding orbitals. σ Electrons are more tightly held and less easily excited. The type of transition can be controlled by the wavelength of light used, since the energy of the light quanta is given by $E = h\nu$.

In the excitation step the electron is promoted with retention of spin to give a singlet excited state. Excited states of higher energy rapidly decay to the lowest excited singlet. This may undergo chemical reaction or it can emit a quantum of light and collapse to the ground state (fluoresce). The singlet excited molecule may also undergo intersystem crossing, with spin inversion, to give a lower energy triplet state. This again may undergo chemical reactions, which are likely to be quite different from those of the singlet, or it may collapse to the ground state with emission of light (phosphoresce). Since this involves spin inversion it is a forbidden process and occurs more slowly than collapse of the excited singlet. The lifetime of an excited singlet is very short since both emission of light and intersystem crossing are fast. (The rate constants for these processes are of the order 10^7–10^8 s^{-1} and 10^8–10^{10} s^{-1} giving lifetimes of 10^{-7}–10^{-8} s and 10^{-8}–10^{-10} s respectively.) The lowest triplet state has a relatively long lifetime ($\sim 10^{-3}$ s). Thus, intramolecular photochemical reactions may involve the singlet excited state, but intermolecular reactions are more likely to involve triplet states since intersystem crossing competes favourably with intermolecular collision.

Both singlet and triplet excited molecules may lose their energy by intermolecular transfer to other molecules. This type of process is utilised in photosensitised reactions where reactant molecules are excited not by direct absorption of light, but through energy transfer from photosensitisers. Light is absorbed by the sensitiser molecule which then transfers its energy to the reactant, if the latter has a lower lying excited state available. If the sensitiser transfers its energy from a singlet state, a singlet excited reactant is produced. If the excited photosensitiser undergoes intersystem crossing to its triplet state and then transfers its energy, a triplet reactant molecule is produced.

Other processes are also possible; for example, triplet* + triplet → singlet* + singlet, as in the generation of singlet oxygen. Thus by careful choice of reaction conditions the chemist can exercise considerable control over the nature of the excited species involved in a photochemical process.

1.3. Experimental investigation of mechanism.[2,4]   The first step in any mechanistic investigation is to determine what all the reactants and products are. This may seem obvious, but incomplete knowledge of all products has often hampered investigations, and has led to wrong conclusions about a mechanism. Minor side products can sometimes provide clues to possible mechanisms.

It must be established that the observed products are actually formed in the reaction studied and not in some subsequent step. Thus, it must be known whether kinetic or thermodynamic control is operative. The fate of particular atoms in a reaction must be known. This is usually evident from the relative positions of substituents in the reactants and products but in more subtle cases atoms can be labelled by isotopes. Detailed stereochemistries of reactants and products must be known since the mechanism has to be able to account for the stereochemical course of the process. For example, we must know whether configuration is retained, inverted or lost at an atom to which a bond is broken and formed.

The second major type of information in a mechanistic investigation comes from rate measurements. Measurement of the rate at varying concentrations of reactants gives an empirical rate equation. In simple cases this gives directly the number of molecules of each type involved in the transition state. In other cases the rate equations are complex and can fit more than one mechanistic scheme. However, for any proposed reaction scheme it is possible to formulate a theoretical rate expression, and this must be compatible with the observed one.

The effects of changing substituents, replacement of atoms by their isotopes, and solvent polarity on the rate of a reaction all give additional useful information. Determination of the rate constant at several different temperatures leads to estimates of enthalpy and entropy of activation.

The information thus obtained can often enable a distinction to be made between concerted and stepwise mechanisms and if the reaction is stepwise it can indicate what sort of intermediate is involved. The following types of experiment have most commonly been used.

Isolation or detection of an intermediate.[5] Stepwise reactions involve intermediates which have widely differing stabilities. At one end of the stability scale these may be isolable; at the other end they may be extremely unstable and their presence only proved indirectly. If an intermediate can be detected by a method which does not affect the mechanism then the stepwise nature of the reaction is proved. The converse does not apply; if no intermediate is detectable the reaction is not necessarily concerted – it may be that the methods of detecting the intermediate are not sufficiently refined.

Sometimes intermediates can actually be isolated. If so, or if a suspected intermediate is available from any other source, it should be subjected to the reaction conditions and shown to undergo the postulated reactions. Intermediates which, although they cannot be isolated, have an appreciable lifetime, may be detected by observation of their infrared, ultraviolet, or nuclear magnetic resonance spectra. Transient radical intermediates can be detected by electron spin resonance spectroscopy provided that the intermediate is present in sufficient concentration ($\sim 10^{-6}$ M).

Certain products which are formed from radical pairs give unusual transient emission lines in their n.m.r. spectra when the reaction is carried out in the n.m.r. probe. Emission lines appear in the same position as the normal absorption lines of the reaction product, and both absorption and emission intensities are often greatly enhanced. They are gradually replaced by the normal absorption peaks. A simple explanation for this phenomenon is that the unpaired electrons of the intermediate readily align with the field. The magnetic field due to these electrons affects the adjacent nuclei, aligning them with or against the applied field. When the radical pair collapses, the product contains a non-equilibrium distribution of nuclear ground states and excited states (nuclear spin states have a longer lifetime than electron spin states, so the nuclear polarisation survives chemical combination). The n.m.r. radiofrequency source can therefore induce emission or absorption of energy to re-establish the normal distribution of nuclear spin states. This effect is known as chemically induced dynamic nuclear polarisation (CIDNP).[6] Examples of the application of the technique are given in chapter 7. A problem with both electron spin resonance and CIDNP, which are very sensitive techniques, is that the radical detected may not be involved in the main reaction but may be part of an independent and minor competing reaction.

Intermediates may also be intercepted by trapping with other reactants. These may be added deliberately to divert the intermediate from its normal mode of breakdown, or they may be solvent molecules or another molecule of starting material. The isolation of side products can often be explained by the involvement of an intermediate and so indicate its presence. Care has to be taken, however, since such side products may be formed in an independent competing reaction and not from branching of an intermediate, as shown. This problem can usually be resolved by a study of the kinetics of the reaction.[5]

reactants ⟶ intermediate ⟨ product A / product B   or   reactants ⟨ product A / product B

Crossover products resulting from intermolecular reaction should always be searched for carefully in any suspected intramolecular reaction. This is done by allowing closely related substrates to react in the presence of each other.

Determination of stereospecificity. Stereospecificity is a widely used criterion of concertedness since it is often easy to determine. A reaction is said to be *stereospecific* if it proceeds by one particular stereochemical course to give only one of several possible stereoisomers. If it merely takes one course predominantly, it is said to be *stereoselective.*

In a concerted reaction, two or more bonds are formed at the same time and since all these bond forming processes are coupled and determined by each other, such reactions are stereospecific (see, for example, the Diels–Alder reaction § 5.1, p. 90).

Stepwise reactions involve intermediates in which bond rotations may be able to compete with the second step of the process. They are therefore often non-stereospecific. For example, in the stepwise radical or ionic 2 + 2 additions of olefins (chapter 6), configuration is lost (1.1).

$$(1.1)$$

Stereospecificity must be used cautiously as a criterion for concertedness. Lack of stereospecificity proves a stepwise mechanism but complete stereospecificity does not prove a concerted mechanism, since other factors, such as steric or polar interactions, may intervene to prevent bond rotations in an intermediate. Alternatively, rapid rotation in an intermediate can result in a stereospecific reaction if one rotamer is highly favoured. Consider as an example a stepwise cycloaddition to an isomeric pair of *cis* and *trans* olefins. If the olefinic substituents are bulky, steric interactions will cause rotamers in which the X groups are *trans* to predominate. Thus, addition to the *trans* olefin may give only the adduct in which the X groups are *trans*. Addition to the *cis* olefin will initially give rotamers with the X groups *cis*. There will then be a driving force for rotation; depending on the relative stabilities of *cis* and *trans* rotamers and on the relative rates of ring closure and rotation, some or all of the product will have the X groups *trans*. In such a case, addition to the *trans* olefin is stereospecific and the lack of stereospecificity is only apparent with the *cis* isomer (1.2.). It is therefore important to test the stereospecificity with both olefins, or at least with one for which rotation is more likely in the intermediate.

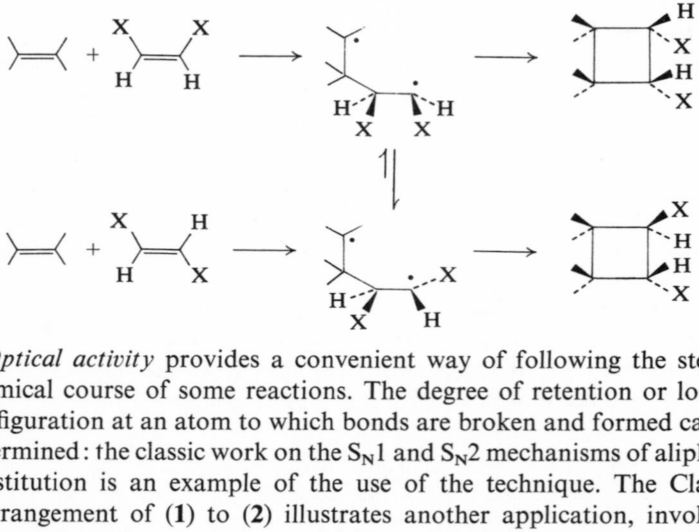

$$(1.2)$$

Optical activity provides a convenient way of following the stereo-chemical course of some reactions. The degree of retention or loss of configuration at an atom to which bonds are broken and formed can be determined: the classic work on the S_N1 and S_N2 mechanisms of aliphatic substitution is an example of the use of the technique. The Claisen rearrangement of (1) to (2) illustrates another application, involving the creation of a new asymmetric centre in the product. This is an example of *asymmetric induction*.

$$(1.3)$$

(1)   (2)

If the reaction is concerted, the C-1 to C-6 bond forms as the carbon–oxygen bond breaks, and complete retention of optical activity should be observed. A stepwise reaction involving cleavage of the carbon–oxygen bond to give a pair of allylic radicals (3) should lead to loss of optical activity, but only if the radical pair has a sufficiently long lifetime for reorientation of the planar radicals to occur.

A concerted mechanism also requires *allylic inversion*; that is, formation of the product in which the migrating allyl group is attached at the carbon which was the terminal one in the starting material. The stepwise mechanism should lead to recombination through both C-1 and C-3 of the allyl radical, so that some of the isomer (**4**) should be produced. Asymmetric induction and allylic inversion are particularly useful in obtaining information about the mechanisms of molecular rearrangements of this type.

(3)   (4)

Determination of substituent effects. The rates of concerted reactions are, in general, less sensitive to changes in substituents than are the rates of stepwise reactions. In a stepwise reaction involving a zwitterionic or diradical intermediate, substituents can stabilise or destabilise the intermediate and therefore profoundly affect the ease of its formation and hence the rate of the overall reaction. The pattern of substituent effects can distinguish between radical and zwitterionic processes since those substituents which stabilise a radical centre do not necessarily stabilise a cationic or anionic centre in the same way.

In a stepwise cycloaddition, substituents can also control the *orientation* (i) or (ii) of the addition (1.4). The product from the more stabilised intermediate predominates because the lower activation energy for its formation means that it is formed faster. Since the orientation of the more stabilised diradical is not always the same as that of the more stabilised zwitterion, the effects of substituents on orientation will often also serve to distinguish between radical and ionic processes.

(1.4)

Concerted reactions where no intermediate is involved are much less sensitive to the nature of substituents. However, since some charge or radical character can be built up by the concerted but unequal formation of bonds, substituent effects can still be significant both as regards the rate and the orientation. Substituent effects are thus a rather indefinite criterion for distinguishing stepwise and concerted processes. Only the trend is general: the greater the effect of change of substituents on the rate, the more likely the reaction is to be stepwise.

There is a very important proviso in the use of substituent effects to study a reaction mechanism, namely that the substituents do not have

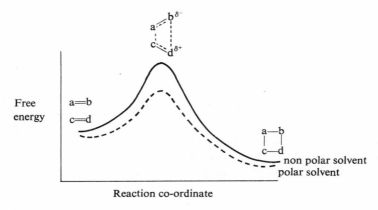

Fig. 1.6

such a profound effect as to completely change the mechanism. As we shall see there are many examples where widely different substituents can change a reaction from a concerted to a stepwise one.

Determination of solvent effects. A reaction going through a zwitter-ionic intermediate generally involves a transition state which is more polar than the reactants and so it should be accelerated by more polar solvents. Solvent polarity is a rather loose term involving dielectric constant and hydrogen bonding ability. The rate of a concerted reaction, or of a stepwise reaction involving a diradical intermediate, where no significant charge build up occurs, should be relatively independent of solvent. This is based on the Hughes–Ingold theory that polar species are better solvated by polar solvents than non-polar species, and are therefore more stabilised. For a concerted reaction in which some asymmetry in the transition state leads to a small charge build up, the energy profiles for the reaction are as shown in fig. 1.6. A moderate solvent effect is to be expected. If the transition state has a higher dipole moment than the reactants the rate will be increased by changing to a more polar solvent. If the transition state has lower overall dipole moment than reactants then the effect of a more polar solvent will be to decrease the rate.

A stepwise reaction via a dipolar intermediate involves two consecutive reactions and the effect of solvent on the activation energies for both reactions must be considered. For example, the effect of change of solvent on the energy profile for the stepwise ionic cycloaddition

$$
\begin{array}{ccc}
a\!=\!b & & a\!-\!b \\
& \longrightarrow & |\quad| \\
c\!=\!d & & c\!-\!d
\end{array}
$$

is illustrated in fig. 1.7.

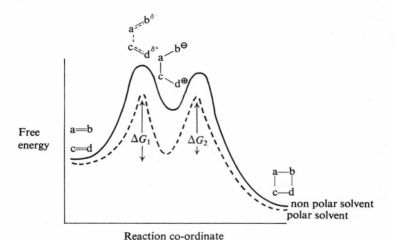

Fig. 1.7

If the first step is rate determining ($\Delta G_1 > \Delta G_2$) then changing from a non polar to a polar solvent lowers ΔG_1 and increases the reaction rate. If the second step is rate determining ($\Delta G_2 > \Delta G_1$) the reverse applies. If $\Delta G_2 \simeq \Delta G_1$, then small increases or decreases in rate occur depending on whether $\Delta G_1 > \Delta G_2$ or $\Delta G_2 > \Delta G_1$. Thus a concerted reaction involving an unsymmetrical transition state and a two-step ionic reaction can only be distinguished when the solvent effect is very large.

Detection of kinetic isotope effects.[7] Isotope effects give information concerning the bonds broken and formed in the rate-determining step of a reaction. They can be classed as primary and secondary isotope effects.

A primary isotope effect is the change observed in the rate of a reaction when one of the atoms of the bond being broken or formed in the rate determining step is replaced by an isotope. The isotope effects most commonly studied in organic chemistry are those resulting from different rates of transfer of hydrogen and deuterium. Consider, as an example, the general reactions:

$$X\text{—}H + Y \rightarrow X...H...Y \rightarrow X + H\text{—}Y$$
$$X\text{—}D + Y \rightarrow X...D...Y \rightarrow X + D\text{—}Y$$

The zero point vibrational energy of X—D will be appreciably lower than that of X—H because of the difference in mass between hydrogen and deuterium. The zero point energies of the transition states for H

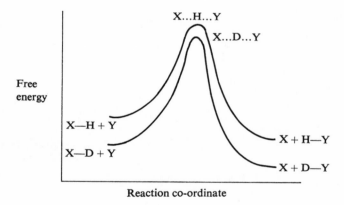

Fig. 1.8

and D transfer will depend on X and Y and on the extent of breaking of X—H (X—D) and formation of H—Y (D—Y). In general X...D...Y will be lower in energy than X...H...Y and the difference in energy between the two states will usually be somewhat less than the energy difference between the reactants (X—H and X—D). The energy profiles for the two reactions will therefore be as in fig. 1.8 and X—H will react faster than X—D.

The magnitude of the isotope effect k_H/k_D will vary with the structure of the transition state. In the limiting case where X = Y and the migrating H is equally bonded to both sites in the transition state, the vibrational energies of X...H...Y and X...D...Y will be almost equal since the main allowed vibrational mode is symmetrical about the central atom and therefore independent of its mass. In such a situation k_H/k_D will be at its maximum value. Hydrogen–deuterium isotope effects are potentially large as isotope effects go because of the large relative difference in mass between the isotopes (this applies even more so to hydrogen–tritium isotope effects). The value of the ratio k_H/k_D ranges from about 1 to 8 or 9 depending on the structure of the transition state. Isotopes other than those of hydrogen can be used although the effects are much smaller; for ^{12}C–^{13}C isotope effects the values range from 1.02 to 1.10. The magnitude of isotope effects depends on the temperature at which they are measured, the values decreasing with increasing temperature. Thus, for example $k_H/k_D = e^{\Delta\Delta G/RT}$ where $\Delta\Delta G$ is the difference in free energy of activation for the reaction of deuterated and undeuterated species. Application of primary hydrogen–deuterium isotope effects is illustrated by the reaction (1.5).

$$(1.5)$$

The deuterated diene rearranges more slowly than the undeuterated one ($k_H/k_D = 5$ at 470°K) indicating that the C—D bond is broken in the rate determining step. A value of 5 for k_H/k_D at 470°K corresponds to a value of 12.2 at 298°K. This exceptionally large value for k_H/k_D indicates that the transition state is highly symmetrical.

Primary isotope effects are also observed in reactions involving simple cleavage or formation of one bond (as opposed to reactions involving transfer of the isotopically labelled atom between two sites). An example is given for the retro Diels–Alder reaction (§ 5.3).

Secondary isotope effects[4] are observed when atoms other than those directly involved in the bond which breaks are replaced by their isotopes. They give basically similar information to primary isotope effects. They can be subdivided into α, β, ... secondary isotope effects depending on whether the isotopes are attached to the atom undergoing bond breaking, or the one adjacent to it, and so on. The effects again arise because the modes of vibration of the reactant and the transition state are different and their zero point vibrational energies are affected to different extents by isotopic substitution: they can be considered simply as reflecting changes in hybridisation. Secondary isotope effects are smaller than primary isotope effects and, depending on how the vibrational modes change on passing from reactants to transition state, k_H/k_D can be greater or less than 1. The former type are known as normal and the latter as inverse secondary isotope effects. An example of their application is to the Diels–Alder reaction, (1.6), which involves the formation of two new bonds.

$$\text{(1.6)}$$

Replacement of any of the hydrogen atoms a, b, c or d by deuterium results in an increase in the reaction rate $(k_H/k_D \simeq 0.9)$, indicating that formation of both new σ bonds occurs simultaneously: there is a simultaneous change of hybridisation at all four termini. In a stepwise reaction, formation of one of the bonds only would be rate determining, so that only isotopic substitution on those atoms undergoing change of hybridisation would be expected to have any effect.

Measurement of activation parameters. The entropy of activation $\Delta S^{\ddagger}$ reflects the difference in ordering between reactants and transition state in the rate determining step of a reaction. This includes not only reactant molecules but also solvent molecules. In a concerted cyclo-addition such as the Diels–Alder reaction the reactants have to be highly orientated with respect to each other. A concerted intramolecular reaction such as the Cope rearrangement also requires a critical orientation of atoms. Such reactions show large negative $\Delta S^{\ddagger}$ values. In general, concerted cycloadditions and rearrangements show large negative entropies of activation. Stepwise processes in general do not require such critical alignment, and smaller activation entropies are to be expected. This tends to be the case, although this criterion must be used cautiously. For example, a concerted cycloaddition must have a large negative $\Delta S^{\ddagger}$, but a cycloaddition with a large negative $\Delta S^{\ddagger}$ need not necessarily be concerted. Concerted and radical reactions do not greatly perturb the ordering of the solvent so that ΔS^{+} is mainly determined by the orientational requirements of the reactants. However, solvent structure is likely to be greatly affected in a reaction proceeding through a highly solvated zwitterion intermediate. Stepwise ionic reactions therefore frequently show large negative ΔS^{+} values.

Not all types of concerted reactions show large negative ΔS^{+}. In a concerted retro cycloaddition, for example, the atoms are restricted similarly in both reactant and transition state. In the latter the σ bonds which break are slightly lengthened compared to those in the reactant. Such a process will therefore be expected to have a ΔS^{+} value of almost zero. In a stepwise retro cycloaddition where only one bond breaks to give an open chain intermediate the ordering of the system is relaxed on going to the transition state which leads to the intermediate, so that a positive entropy of activation is expected.

The degree of ordering in the transition state has also been related to the *volume of activation*. This is based on the principle that the more ordered the transition state, the less volume it should occupy compared with the reactants from which it is derived. For a reaction taking place in solution, the activation volume can be obtained experimentally from a study of the effect of applied pressure on the reaction rate. If the rate increases with pressure, this is taken to indicate that the transition state occupies a smaller volume than the reactants, and if it decreases with pressure, that it occupies a larger volume.

The extrapolation has occasionally been made that an increase of rate with pressure also indicates a concerted process, and a decrease, a stepwise process. This is a dubious assumption, because there is no absolute standard by which the volumes of the transition states for stepwise and concerted processes can be compared.

Concerted processes are usually characterised by having a small enthalpy of activation, ΔH^{+}. Again, however, the converse is not necessarily true: a low ΔH^{+} does not prove a concerted mechanism.

In conclusion, to prove whether a reaction is concerted or stepwise is not a simple matter, since, with the exception of isotope effects, the evidence for a concerted reaction is of a negative rather than a positive nature. The total evidence has to be considered. Frequently, reactions are assumed to be concerted on the basis of one criterion alone, such as stereospecificity. Although this may be justified in a series of closely related reactions where the concerted nature for some has been proved, such incomplete evidence should always be treated with reserve.

REFERENCES

1. R. B. Woodward and R. Hoffmann, 'The Conservation of Orbital Symmetry', *Angew. Chem. Int. Edn*, **8**, 781 (1969).
2. For comprehensive discussions see (*a*) A. A. Frost and R. G. Pearson, *Kinetics and Mechanism*, 2nd edn, Wiley, 1961; (*b*) J. E. Leffler and E. Grunwald, *Rates and Equilibria of Organic Reactions*, Wiley, 1963.
3. Review: A. A. Lamola and N. J. Turro, *Techniques of Organic Chemistry*, vol. xiv, ed. A. Weissberger, Interscience, 1969.
4. For a full discussion of all aspects see S. L. Friess, E. S. Lewis, and A. Weissberger, *Investigation of Rates and Mechanisms of Reactions*, vol. viii of *Techniques of Organic Chemistry*, ed. A. Weissberger, 2nd edn, Interscience, 1961.
5. For a review on the detection of reactive intermediates from kinetic evidence, see R. Huisgen, *Angew. Chem. Int. Edn*, **9**, 751 (1970).
6. The absorption–emission pattern of the CIDNP spectrum can give valuable information concerning the spin multiplicity of the radical pair precursors: R. Kaptein, *Chem. Comm.*, 1971, 732.
7. Reviews: L. Melander, *Isotope Effects on Reaction Rates*, Ronald Press, 1960; W. H. Saunders, *Survey Progress Chem.*, **3**, 109 (1966); see also ref.4, p.389.

2 Theory of concerted reactions

The development of a general theory of concerted reactions has been due chiefly to the work of R. B. Woodward and R. Hoffmann.[1] They have taken the basic ideas of molecular orbital theory and used them, mainly in a qualitative way, to derive selection rules which predict the stereochemical course of various types of concerted reactions. These rules are best understood in terms of symmetries of interacting molecular orbitals. However, it is important to appreciate that the rules do not depend on the use of any particular bonding theory for their validity. The same selection rules have been obtained by various other theoretical approaches, including a valence bond treatment,[2] and several alternative descriptions of the rules are commonly used in the chemical literature. In this chapter the aim is to describe the most important of these theoretical approaches and to show, as far as possible, how they are interrelated.

2.1. Molecular orbitals. Since extensive use is made of orbital pictures in describing the course of concerted reactions, the basis of these orbital representations will be outlined.

Molecular orbitals are made up of combinations of atomic orbitals. Each molecular orbital can contain a maximum of two electrons. The interaction of two equivalent atomic orbitals leads to the formation of two new molecular orbitals, one of lower energy than the atomic orbitals (the bonding orbital) and the other of higher energy (the antibonding orbital). The positive overlap between the atoms in the bonding orbital can be represented by signs: the $+$ signs in the larger lobes of the C—C σ bond illustrated (fig. 2.1) show this bonding interaction. For the σ^* bond, however, the lack of overlap is indicated by the sign discontinuity between the atoms. This is a formalised way of indicating a change of phase in the wave function.

σ (C—C)   σ* (C—C)

Fig. 2.1

Here the shapes of the interacting lobes represent those of the hybrid orbitals on the individual atoms, and are not drawn to represent areas of electron density in the resulting C—C bond. The important feature of this type of bond is that the electron density is concentrated along the axis joining the nuclei.

An important difference between the bonding and antibonding orbitals is their *symmetry*: the C—C σ orbital is symmetric (S) with respect to a plane perpendicular to the axis of the bond, but the σ* orbital is antisymmetric (A).

The same formalised system can be used to represent a C—H σ bond; here the interacting orbitals are a hybrid orbital on carbon and an s-orbital on hydrogen (fig. 2.2).

σ (C—H)   σ* (C—H)

Fig. 2.2

In a similar way delocalised molecular orbitals can be constructed from a system of equivalent overlapping p-atomic orbitals. The combination of n such p-orbitals will lead to the formation of n molecular orbitals, symmetrically distributed in energy about the non-bonding level. Thus, for example, two such p-orbitals can combine to give a π bonding orbital and a π* antibonding orbital as in ethylene (fig. 2.3). As before, equivalence of signs indicates positive overlap and bonding interaction.

π (C=C)   π* (C=C)

Fig. 2.3

Again, the different symmetries of the two orbitals distinguish them. All planar π systems have a plane of symmetry (m_1) bisecting the p-orbitals (the nodal plane) about which they are antisymmetric. There are, however, other symmetry elements in the ethylene π systems, including another mirror plane, m_2, perpendicular to the C—C bond, and a twofold axis, C_2, running through the C—C bond and perpendicular to it (fig. 2.4).

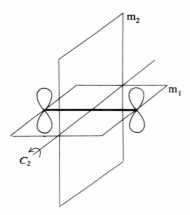

Fig. 2.4

The π orbital is symmetric with respect to m_2 and π^* is antisymmetric. On the other hand, if the twofold axis is taken as the element of symmetry, π is antisymmetric and π^* is symmetric. Thus, the element of symmetry must be specified when orbitals are classified as symmetric or antisymmetric.

The overlap of three p-orbitals gives the allyl system; the three molecular orbitals constructed from them are shown in fig. 2.5. The orbitals are numbered in order of increasing energy. As the energy increases, the number of sign discontinuities or nodes increases. ψ_2 has no sign on the central carbon because the wave function goes through a node at this point. The same three symmetry elements are present as were specified for the ethylene π system; the classification of the orbitals as symmetric (S) or antisymmetric (A) with respect to the mirror plane m_2 and the twofold axis is as shown.

In the ground state, electrons occupy these orbitals, two in each, from the lowest upwards. Hence the highest occupied π orbital can be determined. For the allyl cation, with two π electrons, it is ψ_1; for the allyl radical and anion with respectively three and four π electrons, it is ψ_2. In general, odd electron systems correspond to the even electron systems which have one more electron.

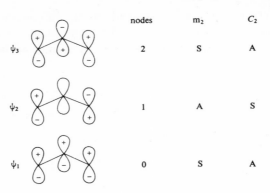

	nodes	m_2	C_2
ψ_3	2	S	A
ψ_2	1	A	S
ψ_1	0	S	A

Fig. 2.5

For butadiene in the *cisoid* conformation the orbitals and their symmetry classifications with respect to m_2 and C_2 are as shown in fig. 2.6. Neither of these symmetry elements is present in the *transoid* conformation; with polyenes of more than three atoms, therefore, the geometry must also be specified.

In general, for linear all *cis* polyenes with n atoms, the n π orbitals can be drawn by putting in the signs with zero nodes for ψ_1, one node for ψ_2, and so on, up to $n-1$ nodes for ψ_n. For polyenes in their ground

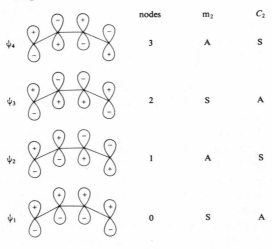

	nodes	m_2	C_2
ψ_4	3	A	S
ψ_3	2	S	A
ψ_2	1	A	S
ψ_1	0	S	A

Fig. 2.6

states, the highest occupied molecular orbital (HOMO) will be symmetric with respect to m_2 for 2, 6, 10,... electron systems, and antisymmetric for 4, 8, 12,... electron systems. The lowest vacant orbital (LVMO) always has the opposite symmetry to that of the HOMO.

2.2. Frontier orbital approach. The first moves towards a theory of concerted reactions came in interpretations of the results of thermal and photochemical cyclisations of polyenes (electrocyclic reactions, chapter 3), which were observed to be stereospecific, the direction of ring closure depending on whether the reaction was brought about by heat or by ultraviolet irradiation. In a study of thermal and photochemical isomerisations of compounds related to calciferol (vitamin D, **1**), Havinga and Schlatmann remarked on the striking stereochemical difference between thermal and photo induced ring closure.[3] Calciferol thermally tautomerises to precalciferol (**2**). At 150–200°C (**2**) then cyclises stereospecifically to give two *cis* fusion products, pyrocalciferol (**3**) and

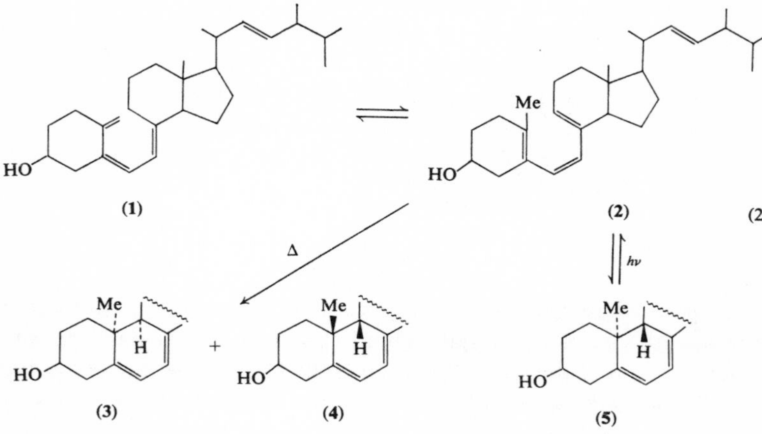

(1)   (2)   (2.1)

(3)   (4)   (5)

isopyrocalciferol (**4**). If precalciferol is irradiated, however, it equilibrates with a different ring closed isomer, the *trans* fused lumisterol (**5**). These relationships are summarised in equation 2.1.

To reduce these results to the simplest terms, a triene is observed to cyclise thermally to give, stereospecifically, products in which the groups X and Y are *cis*, and photochemically to give a product where X and Y are *trans* (2.2).

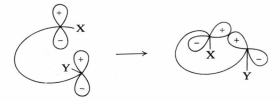

$$(2.2)$$

This work was published in 1961. In the paper the authors quote a suggestion by Oosterhoff that the symmetry of the HOMO of the triene might control the direction of ring closure. The HOMO of a hexatriene, a six π electron system, is symmetric with respect to a mirror plane; thus, the terminal lobes are in phase. Movement of these lobes to provide a bonding interaction in the transition state should lead to *cis* alignment of the substituents X and Y (fig. 2.7).

Fig. 2.7

For a photochemical reaction an electron is assumed to be promoted to the lowest vacant molecular orbital (LVMO), which has opposite symmetry to that of the ground state HOMO. In this case, therefore, the terminal lobes will be out of phase, and the rotation to give bonding overlap will lead to *trans* alignment of the groups X and Y (fig. 2.8).

These ideas were not fully developed until 1965, when Woodward and Hoffmann constructed a general theory of electrocyclic reactions. This resulted from some very similar observations on the direction of cyclisation of trienes, made during work connected with the synthesis of vitamin B_{12}.[4] This theory is expounded in chapter 3; however, the use of the HOMO to predict the stereochemistry of the reaction needs some explanation. It seems intuitively right, since the electrons of the HOMO can be likened to the valence electrons of an atom; that is, they

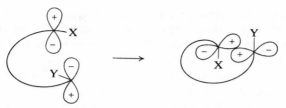

Fig. 2.8

are the ones most likely to control the mode of reaction of the system. The exclusive use of the HOMO must be an oversimplification, since all the occupied orbitals contribute to some extent to overlap stabilisation in the transition state.

This use of the HOMO to predict the course of ring closure is an extension to unimolecular reactions of *frontier orbital theory*, which has more commonly been used to explain the course of bimolecular reactions.[5] The interaction between reagent and substrate (a conjugated π system, for example) takes place by overlap of filled orbitals of the substrate with empty orbitals of the reagent, and vice versa (fig. 2.9). The orbitals must be of like symmetry to interact. The result of the interaction is stabilisation of the filled orbital. The strength of the

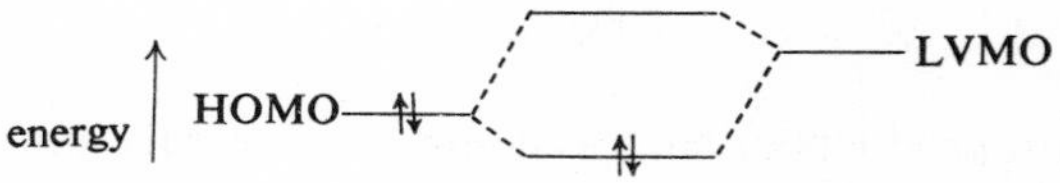

energy | HOMO LVMO

Fig. 2.9

interaction is greater, the closer the interacting filled and empty orbitals are in energy; thus the most important contributions are likely to be from the interaction of the HOMO of one component with the LVMO of the other.

The new orbitals produced by the interaction can be regarded as 'transition state' orbitals. For bimolecular reactions involving the interaction of two π systems, the predictions of the theory can be simply stated as follows: if the symmetry of the HOMO of one component is such that it can overlap with the LVMO of the other component, then the reaction is favoured as a concerted process. It is said to be *symmetry-allowed*. If the orbitals are of the wrong symmetry for overlap, then the reaction is *symmetry-forbidden*.

It is important to appreciate that these terms refer only to *concerted* processes. By choosing an appropriate transition state geometry, it is possible to devise a symmetry-allowed concerted mechanism for a great many reactions, but this does not mean that all these reactions are concerted. If the geometrical constraints in the transition state are severe, the reactants will probably choose an alternative stepwise pathway of lower energy. When the transition state for an allowed concerted process is not strained, then the reaction will go predominantly by the concerted pathway. Even in these cases, a stepwise pathway may not be much higher in energy than the concerted route. Careful investigation

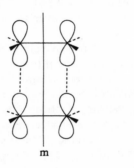 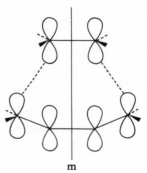

Fig. 2.10

of the reaction products of such systems often reveals minor products which must be formed by stepwise mechanisms competing with the major, concerted mechanism. On the other hand, if geometrical constraints rule out all possible symmetry-allowed mechanisms, the system will go by a stepwise route rather than by a symmetry-forbidden concerted route. The rules therefore indicate only which geometries of approach are feasible for concerted reaction, and do not exclude the possibility that a stepwise route may be of lower energy than any concerted route.

The frontier orbital theory has its critics,[6] and its assumptions are indeed rather drastic, but for simple bimolecular concerted reactions its predictions are borne out in practice. Consider, as examples, the approach of two ethylene molecules, and of ethylene and butadiene, in the manner shown in fig. 2.10.

Each transition state has a mirror plane (m). In the ground state, the occupied π orbital of ethylene is symmetric with respect to this plane, and the unoccupied π^* orbital is antisymmetric. For butadiene the HOMO is antisymmetric and the LVMO is symmetric. Thus, when two ethylene molecules approach each other the occupied and vacant orbitals are of the wrong symmetry to interact and the reaction is symmetry-forbidden. For ethylene and butadiene, the LVMO of either partner is of the correct symmetry to interact with the HOMO of the other, and the reaction is symmetry-allowed (fig. 2.11).

Note particularly that the *geometry of approach* is a key factor in determining whether a reaction is allowed or forbidden. If the components approach each other in a different way to that illustrated,

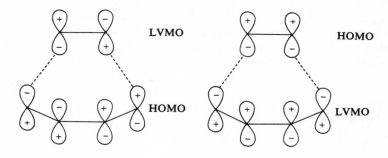

Fig. 2.11

different symmetry elements will be present in the transition state, and the HOMO and LVMO may have different symmetries with respect to these new elements.

For photochemical reactions the predictions are the reverse of those for thermal reactions. In a bimolecular reaction it is assumed that an electron is excited from the HOMO to the LVMO of one component, the other component remaining in its ground state. The new HOMO of the excited component now has the opposite symmetry to that of the ground state species, so interactions with the other component which were forbidden in the ground state become allowed, and vice versa. This simple approach does not tackle such problems as whether the first excited state is indeed the one from which reaction takes place, or whether the spin state (singlet or triplet) is relevant. In practice however, it is remarkable how often its basic prediction is borne out – that reactions which are thermally forbidden can be brought about photochemically, and vice versa.

2.3. Correlation diagrams. Correlation diagrams, which have been used extensively by Woodward and Hoffmann in their development of the theory of concerted reactions, can be regarded as an extension of the frontier orbital approach. They provide a means of following the energy changes of all the participating electrons in a concerted reaction, and not only those in the HOMO. Relevant unoccupied orbitals are also included.

Correlation diagrams are not new; they were used in the early days of the development of molecular orbital theory to follow energy changes of the orbitals of separate atoms as they approach through space to form a molecule. Their use is discussed, for example, by Coulson.[7] Longuet-Higgins and Abrahamson showed that correlation diagrams could be applied to follow the energy changes in the orbitals of molecules as they approached through space to form a 'united molecule', and successfully explained the course of several concerted reactions by this method.[8]

$$C_2 \qquad\qquad\qquad\qquad\qquad\qquad\qquad m \qquad\quad (2.3)$$
$$\text{conrotatory} \qquad\qquad\qquad\qquad\qquad\qquad \text{disrotatory}$$

Symmetry plays a vital role in the construction of these diagrams. The symmetry elements present in the transition state are defined, and the participating orbitals, and those of the product, classified as symmetric or antisymmetric with respect to each of these elements. Each orbital is then transformed, with *conservation of symmetry*, into an orbital of the product. Substituents which do not fundamentally alter the energies of the participating orbitals, such as alkyl groups, are ignored for the purposes of determining the symmetry.

As an example of the use of correlation diagrams, consider the closure of butadiene to cyclobutene, and the reverse process (the conditions needed to bring about these reactions are discussed in chapter 3). The reaction involves rotation of the terminal CH_2 groups of the diene through 90°. The groups can rotate in the same direction (*conrotatory* movement) or in opposite directions (*disrotatory* movement); these are shown in equation 2.3.

Although the products of the two types of ring closure are not distinguishable in this case, they may be with substituted butadienes. The conrotatory mode maintains a twofold axis of symmetry (C_2) whereas in the disrotatory closure a mirror plane (m) is maintained. The symmetry classification of the participating orbitals depends on which mode of closure is used.

Having considered the various possible geometries for the transition state and the symmetry associated with each, the next step is to note which bonds are made and which are broken in the reaction, and to list the molecular orbitals associated with these bonds. In the closure of butadiene to cyclobutene, the π bonds of the diene are broken; those formed are the π bond between C-2 and C-3 and the σ bond between C-1 and C-4. The carbon σ framework is otherwise basically unchanged during the reaction and the orbitals associated with it are not included in the correlation diagram. Similarly the orbitals associated with the C—H bonds are not included because changes in their energies are small compared with the other bonding changes.

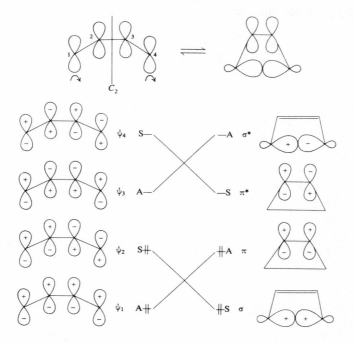

Fig. 2.12

The orbitals associated with the bonds made and broken are:

butadiene: $\psi_1, \psi_2, \psi_3, \psi_4$
cyclobutene: $\sigma, \pi, \pi^*, \sigma^*$.

Next, for each of the possible transition state geometries, the orbitals are classified (S or A) with respect to the appropriate symmetry element. The symmetry elements chosen must bisect bonds made or broken during the reaction. For the conrotatory mode of closure, where a two-fold axis is maintained, the symmetries are as shown (fig. 2.12).

The joining of orbitals of like symmetry completes the correlation diagram: thus, ψ_1 of butadiene correlates with π of cyclobutene, and ψ_2 with σ. The four electrons occupying ψ_1 and ψ_2 in the ground state of butadiene can therefore go into the π and σ bonding orbitals of cyclobutene without involving a high energy transition state: the reaction is symmetry-allowed for the conrotatory mode. Similar correlations exist for the antibonding orbitals. The important characteristic of a correlation diagram of an allowed reaction is this linking of bonding with bonding orbitals, and antibonding with antibonding orbitals.

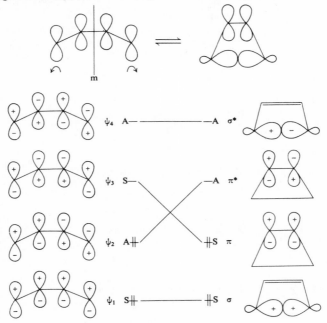

Fig. 2.13

For the disrotatory mode of closure, the relationship of orbitals is different. The symmetries for the disrotatory closure are as shown in fig. 2.13.

The orbital ψ_2, occupied in the ground state of butadiene, cannot correlate with either of the bonding orbitals of the product, and it increases sharply in energy as the transition state is approached; the reaction is therefore symmetry-disallowed.

The same procedure is used in the construction of correlation diagrams for bimolecular reactions. In some cases it is not a simple matter to assign symmetries to all the participating orbitals. Consider, for example, a cycloaddition reaction in which two new σ bonds are formed, a plane of symmetry (m) being maintained throughout the reaction (fig. 2.14). The orbitals σ_1 and σ_2, localised on the new σ bonds, have no symmetry with respect to this plane.

However, by taking linear combinations of σ_1 and σ_2, two new σ bonding orbitals can be produced, having symmetry with respect to the mirror plane. The new orbitals are $(\sigma_1 + \sigma_2)$, which is symmetric with respect to m, and $(\sigma_1 - \sigma_2)$, which is antisymmetric (fig. 2.15). Each of these new σ bonding orbitals, like any other molecular orbital, can contain a maximum of two electrons.

Combinations of the antibonding orbitals can be used in a similar way. This device enables correlation diagrams to be constructed for a wide range of reactions where there is symmetry in the transition state.

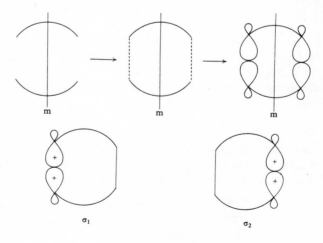

Fig. 2.14

There are some difficulties in the general application of correlation diagrams to concerted reactions. One is that many transition states have no symmetry, even when non-reacting substituents are ignored. Another is that it is difficult to define the point at which the effect of substituents can no longer be neglected: for example, if one of the components in a Diels–Alder cycloaddition is highly polarised, because of the nature of its substituents or because its skeleton contains a hetero-atom, it is probable that the transition state for the reaction will be asymmetric,

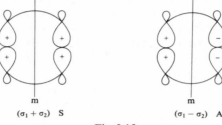

Fig. 2.15

the formation of one σ bond being much more advanced than of the other. The simple ethylene–butadiene correlation diagram might then be misleading. The application of correlation diagrams to explain the course of photochemical reactions also needs great caution (chapter 3). Correlation diagrams do, however, succeed in giving a good insight into the reasons why some reactions are thermally favoured and others are not.

The correlation diagrams that have been constructed so far involve relationships of orbitals; another type, *state correlation diagrams*, link states of like symmetry. The state symmetry for a molecule is obtained by multiplying together the symmetry labels for each electron, using the rules:

$$S \times S = S; \quad A \times A = S; \quad S \times A = A; \quad A \times S = A.$$

Consider again the example of the conrotatory and disrotatory closure of butadiene to cyclobutene. The symmetry classifications of the orbitals involved, with respect to the mirror plane and the twofold axis, are:

Twofold axis	$S: \psi_2, \psi_4, \sigma, \pi^*$;
(conrotatory)	$A: \psi_1, \psi_3, \pi, \sigma^*$.
Mirror plane	$S: \psi_1, \psi_3, \sigma, \pi$;
(disrotatory)	$A: \psi_2, \psi_4, \pi^*, \sigma^*$.

The ground state of butadiene ($\psi_1^2\psi_2^2$) is (AA)(SS) = S with respect to a twofold axis. The first excited state ($\psi_1^2\psi_2\psi_3$) is (AA)(S)(A) = A. Similarly state symmetries for other excited states of butadiene and for the states of cyclobutene can be calculated, and a state correlation diagram drawn (fig. 2.16). The states which are linked have matching sets of symmetry labels for the electrons, as well as having the same overall state symmetry. The broken lines indicate the actual paths followed in the interconversion of states of A symmetry.

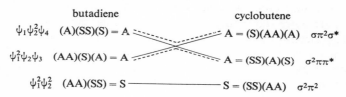

butadiene   cyclobutene

$\psi_1\psi_2^2\psi_4$ (A)(SS)(S) = A $\cdots$ $A = $ (S)(AA)(A) $\sigma\pi^2\sigma*$

$\psi_1^2\psi_2\psi_3$ (AA)(S)(A) = A $\cdots$ $A = $ (SS)(A)(S) $\sigma^2\pi\pi*$

$\psi_1^2\psi_2^2$ (AA)(SS) = S ———————— S = (SS)(AA) $\sigma^2\pi^2$

Fig. 2.16. State correlation diagram (conrotatory).

Because of the rule that states of like symmetry cannot cross, the actual correlations of the A states will be those shown by the dotted lines. Fig. 2.16 shows that the conrotatory mode of closure is allowed for the ground state but not for the first excited state. A similar state correlation diagram for the disrotatory closure (fig. 2.17) reveals the energy barrier to the ground state reaction.

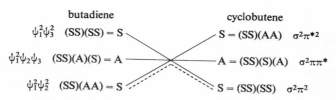

butadiene   cyclobutene

$\psi_1^2\psi_3^2$ (SS)(SS) = S $\quad$ $S = $ (SS)(AA) $\sigma^2\pi*^2$

$\psi_1^2\psi_2\psi_3$ (SS)(A)(S) = A ———— $A = $ (SS)(S)(A) $\sigma^2\pi\pi*$

$\psi_1^2\psi_2^2$ (SS)(AA) = S $\quad$ $S = $ (SS)(SS) $\sigma^2\pi^2$

Fig. 2.17. State correlation diagram (disrotatory).

State correlation diagrams are particularly useful for predicting the course of reactions from excited states; an initial increase in energy, as indicated by the upward slope of the line from a given state, is a sign that the reaction from that state has a high energy barrier.

From the results obtained by the application of these concepts of symmetry to a wide range of concerted reactions, Woodward and Hoffmann have been able to formulate a generalised selection rule for predicting whether or not a reaction is symmetry-allowed. Equivalent forms of the same rule have been arrived at independently by Zimmerman[9] and by Dewar.[6] Since their approaches to the theory are different from that of Woodward and Hoffmann, these will be described first, and the Woodward–Hoffmann rule then compared with the alternative statements.

2.4. The 'aromatic transition state' concept. One of the major concepts currently used in organic chemistry is that of aromaticity.[10] The Hückel rule, that planar cyclic polyenes with $(4n + 2)$ π electrons (n being an integer) are more stable than their acyclic counterparts whereas those with $4n$ π electrons are less stable, is supported both by theory and by practical experience. This striking alternation in the stabilities of cyclic polyenes suggests a possible connection with the equally striking alternation in the stabilities of cyclic transition states of concerted

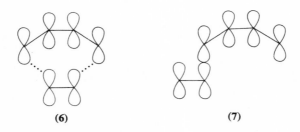

(6)   (7)

reactions. This is the basis of the theories of both Dewar and Zimmerman; just as the $(4n + 2)$ π cyclic polyenes can be described as *aromatic* and the $4n$ π systems as *antiaromatic*, so the cyclic transition states of concerted reactions can be classified as aromatic or antiaromatic. To take a simple example, the thermal $2\pi + 2\pi$ cyclodimerisation of olefins prefers to go by a stepwise mechanism, whereas the $4\pi + 2\pi$ Diels–Alder cycloaddition is generally concerted. It may therefore be concluded that the four-electron cyclic transition state is destabilised relative to its acyclic counterpart whereas the six-electron cyclic transition state is more stable than its acyclic counterpart. To illustrate the point, the cyclic transition state (6) for the Diels–Alder reaction has an orbital arrangement equivalent to that of benzene and the acyclic transition state (7), one like that of hexatriene. The cyclic transition state should be more stable for the same reasons that benzene is more stable than hexatriene.

This concept is extremely useful in practice. In order to make full use of the concept it is necessary to introduce another idea, that of Möbius systems. If a planar, linear polyene is twisted so that one end is turned through 180° relative to the other, and the ends of the polyene are then joined, the top portion of the π system will overlap with what was the bottom, and the molecule will be a so-called Möbius polyene. The topology of the π system is that of a Möbius strip: a simple model can be made by taking a narrow strip of paper, twisting one end through 180°, and then joining the ends together. The Möbius strip is remarkable in that it has a single continuous surface; similarly the p-orbitals of the Möbius polyene form a single continuous ring (fig. 2.18) instead of the two separate rings of the normal cyclic π system.

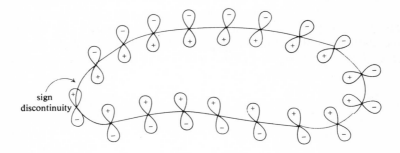

Fig. 2.18. Möbius π system.

In theory one can imagine similar cyclic polyenes with 2, 3, 4 ... twists. Those with zero or an even number of twists are classed as Hückel polyenes, and those with an odd number of twists are classed as Möbius polyenes. Twisted polyenes of this type are as yet unknown, and will presumably be highly strained, reactive species if they are ever made. On the other hand the possibility of having twisted Hückel or Möbius *transition states* in pericyclic reactions is a real one.

The predictions of molecular orbital theory for Hückel cyclic polyenes have already been stated: they are stabilised relative to their acyclic counterparts when they contain $(4n + 2)$ π electrons and destabilised when they contain $4n$ π electrons. For the Möbius systems the predictions are just the opposite. Möbius polyenes should be stabilised when they contain $4n$ π electrons and destabilised when they contain $(4n + 2)$ π electrons. Thus, if both types of cyclic transition state are obtainable, the predictions for thermal pericyclic reactions can be summarised as follows:

number of electrons	*Hückel type*	*Möbius type*
0, 4, 8, ..., $4n$	unfavourable	favourable
2, 6, 10, ..., $(4n + 2)$	favourable	unfavourable

For photochemical reactions the rules are reversed.

As an example of a reaction which can be regarded as having a Möbius transition state, consider the conrotatory ring closure of butadiene described earlier. The signs on the p-lobes of the butadiene are chosen to minimise the number of out of phase overlaps (fig. 2.19). This choice is an arbitrary one and is made for simplicity; the reversal of the sign on one of the lobes would just introduce two more sign inversions into the transition state, and leave its classification (as a Hückel or Möbius system) unaltered. The important point is that the method does not depend on knowing the symmetries of the various molecular orbitals of the polyene, and so can be applied to transition states with no symmetry.

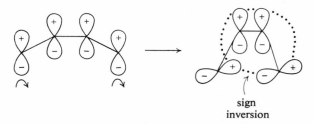

Fig. 2.19

In this reaction the transition state is a Möbius system because there is a sign inversion between the overlapping p-lobes of C-1 and C-4, and since four electrons are involved, the reaction should be thermally favourable.

If the $4n$ π Möbius systems are included in the definition of aromatic compounds, the general rule for thermal pericyclic reactions can be succinctly stated as follows:

'*Thermal pericyclic reactions take place via aromatic transition states.*'

Although the basic idea of this rule was actually first proposed in 1939 by Evans,[11] its revival and justification is mainly due to Dewar.

It is probably the simplest and clearest statement of the requirements for a concerted reaction to go thermally via a cyclic transition state. The procedure for using it is as follows:

1. Draw the transition state as a series of overlapping s- and p-orbitals, putting in + and − signs so that they minimise the number of sign changes in the participating orbitals.
2. Count the number of sign inversions in the cyclic array and the number of electrons involved. If the cycle includes both lobes of a single p-orbital as in fig. 2.19, the necessary sign inversion across these two lobes is not counted.
3. Classify the transition state as being of the Hückel type (zero or an even number of sign inversions) or of the Möbius type (odd number of sign inversions). From the number of electrons determine whether or not it is aromatic.

Examples of the use of the rule are given extensively in later chapters, but a few illustrations here may help to clarify the procedure.

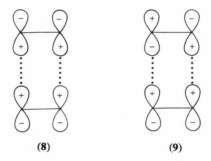

(8)   (9)

Cyclodimerisation of ethylene. The simplest transition state (8) for the cyclodimerisation is shown, the signs chosen indicating minimum phase change in the participating components. The four carbon atoms are coplanar, with the top lobes of one π system interacting with the bottom lobes of the other. The transition state is of the Hückel type and as it contains four electrons it is antiaromatic. The reaction is therefore not favoured thermally. Note that the choice of signs for the lobes is arbitrary: if one is reversed, as in (9), the transition state now has two sign inversions but is still of the Hückel type.

Opening of a cyclopropyl cation. This reaction (fig. 2.20), like the cyclisation of butadiene, is an electrocyclic process, and as such, can be conrotatory or disrotatory. For the conrotatory opening the transition state (**10**) is of the Möbius type; for the disrotatory opening it is of the Hückel type (**11**). Since two electrons are involved, the opening should therefore be disrotatory.

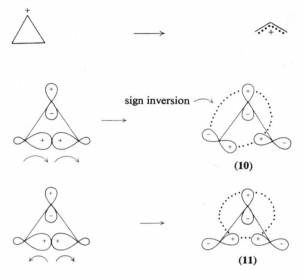

Fig. 2.20

1,3-Migration of a substituent through an allyl system. In this reaction (fig. 2.21) a group R migrates across the top face of an allyl system from C-1 to C-3.

Fig. 2.21

Two transition states are possible, depending on whether R uses an s-orbital (**12**) or a p-orbital (**13**) for bridging.

Transition state (**12**) is of the Hückel type and (**13**) is of the Möbius type. Since four electrons participate, (**12**) is not allowed for a thermal reaction but (**13**) is. This prediction is borne out in practice; it has interesting stereochemical consequences, which are discussed in chapter 7.

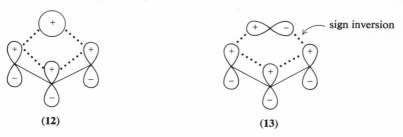

(12)   (13)

2.5. General rule for pericyclic reactions. A general rule for determining the stereochemical course of concerted reactions has been put forward by Woodward and Hoffmann. It requires some preliminary definitions and comments.

Consider a pericyclic reaction in which the electrons of a π bond are used in the transition state, and in which new bonds are being formed at the termini. There are two stereochemically distinct ways in which the new bonds can be formed: either to the same face of the π bond (that is, in a *suprafacial* way), or to opposite faces (that is, in an *antarafacial* way), (fig. 2.22).

suprafacial   antarafacial

Fig. 2.22

The same definitions apply to longer π systems. Similarly, there are stereochemically distinct ways in which a σ bond can be used. Suprafacial use of a σ bond involves bonding in the same way at both termini; that is, retention of configuration at both ends, or inversion at both ends. Antarafacial use of a σ bond involves retention at one end and inversion at the other. These are illustrated in fig. 2.23.

A non bonding p-orbital which participates in a pericyclic reaction can form bonds to both the flanking groups from the same lobe (suprafacially) or from opposite lobes (antarafacially) (fig. 2.24).

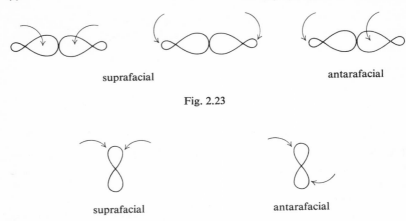

suprafacial   antarafacial

Fig. 2.23

suprafacial   antarafacial

Fig. 2.24

The symbols π, σ, and ω are given respectively to the π systems, σ bonds and lone p-orbitals which participate in the transition state, and the symbols (s) and (a) to indicate their suprafacial or antarafacial use. The notation is completed by the number of electrons supplied by each component. Thus, $_\pi 2_s$ denotes a two-electron π system used in a suprafacial way, $_\omega 0_a$ indicates a vacant p-orbital used in an antarafacial way, and so on.

Woodward and Hoffmann state the general rule for pericyclic reactions as follows:

'A ground state pericyclic change is symmetry-allowed when the total number of $(4q + 2)$ suprafacial and $4r$ antarafacial components is odd' (*q* and *r* being zero or integers).

To apply the rule, an orbital picture of the reactants is constructed and the components used in a geometrically feasible way to achieve overlap. The $(4q + 2)$ electron suprafacial and $4r$ electron antarafacial components are counted, the others being ignored. If the total is an odd number, the reaction is predicted to be thermally allowed.

Although the rule may seem to bear little relationship to the aromatic transition state concept described earlier, it is completely consistent. The use of the same examples as in § 2.4 will illustrate the essential similarity.

Cyclodimerisation of ethylene. The reaction via a planar transition state is a $_\pi 2_s + _\pi 2_s$ reaction, (**14**). There are two $(4q + 2)$ electron suprafacial components and no antarafacial components; the reaction is thermally disallowed, since the total number of 'counting' components is two, an even number.

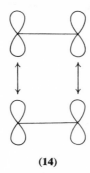

(14)

A comparison of diagrams **(14)** and **(8)** shows that the suprafacial, suprafacial interaction corresponds to a choice of signs for the ethylene components which gives the minimum phase change.

Opening of cyclopropyl cation. The conrotatory ring opening **(15)** is a thermally disallowed $_\sigma 2_a + _\omega 0_s$ reaction. Note the similarity between the overlap here and in diagram **(10)**. The geometrical constraints require that if the σ bond is used in an antarafacial way, then both bonds must be made to the same face of the lone p-orbital. Alternatively, the σ bond could be used in a suprafacial way, but then the overlap must be to opposite faces of the p-lobe, **(16)**; the reaction, now classed as $_\sigma 2_s + _\omega 0_a$, is still thermally disallowed. A disrotatory opening, for example the $_\sigma 2_s + _\omega 0_s$ process illustrated, **(17)**, is thermally allowed.

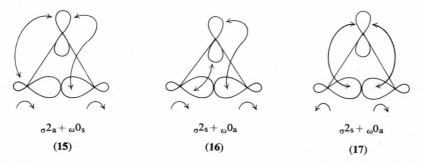

$_\sigma 2_a + _\omega 0_s$ $_\sigma 2_s + _\omega 0_a$ $_\sigma 2_s + _\omega 0_a$

(15)   **(16)**   **(17)**

Migration of a substituent through an allyl system. The reaction could involve retention of configuration at the migrating group, **(18)**, or inversion, (19). The former, illustrated as a $_\sigma 2_s + _\pi 2_s$ process, is thermally disallowed. The latter, shown as a $_\sigma 2_a + _\pi 2_s$ process, is allowed; in the transition state, rehybridisation of the sp^3-orbital to a p-orbital will presumably take place, making this representation (19) equivalent to that of diagram (13).

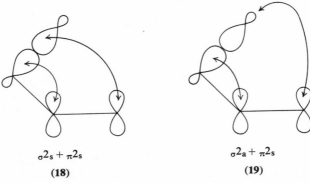

$_\sigma 2_s + _\pi 2_s$ $_\sigma 2_a + _\pi 2_s$

(18)   (19)

This generalised rule of Woodward and Hoffmann can also be linked with the frontier orbital approach to the theory of concerted reactions. The alkyl migration (19) can be represented as an interaction of σ with π^*, (20), or of σ^* with π, (21).[5] In either representation, overlap in the transition state is to the rear face of the alkyl group, as the rules require. Thus, these apparently different approaches to the theory all lead to the same prediction.

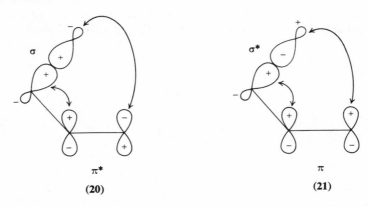

σ   σ*

π*   π

(20)   (21)

In the text, explanations for the stereochemical course of particular reactions have been selected from the various alternatives presented here.

Some approaches seem more suited than the others to particular examples, although in principle, any of the above approaches can be adapted to explain all the different types of concerted processes.

REFERENCES

1. Reviews: R. Hoffmann and R. B. Woodward, 'The Conservation of Orbital Symmetry', *Accounts Chem. Research*, **1**, 17 (1968); R. B. Woodward and R. Hoffmann, *Angew. Chem. Int. Edn*, **8**, 781 (1969).
2. J. J. C. Mulder and L. J. Oosterhoff, *Chem. Comm.*, 1970, 305.
3. E. Havinga and J. L. M. A. Schlatmann, *Tetrahedron*, **16**, 146 (1961).
4. Woodward has described the experimental work which led to the development of the theory of symmetry control in *Aromaticity*, Chem. Soc. Special Publication no. 21, p. 217 (1967).
5. Review: K. Fukui and H. Fujimoto, 'Orbital Symmetry and Electrocyclic Rearrangements', in *Mechanisms of Molecular Migrations*, vol. 2, ed. B. S. Thyagarajan, Interscience, 1969, p. 117. See also L. Salem, 'Orbital Interactions and Reactions Paths', *Chem. in Britain*, **5**, 449 (1969).
6. M. J. S. Dewar, *The Molecular Orbital Theory of Organic Chemistry*, McGraw-Hill, 1969.†
7. C. A. Coulson, *Valence*, 2nd edn, Oxford, 1961.
8. H. C. Longuet-Higgins and E. W. Abrahamson, *J. Amer. Chem. Soc.*, **87**, 2045 (1965).
9. H. E. Zimmerman, *J. Amer. Chem. Soc.* **88**, 1564, 1566 (1966); *Angew. Chem. Int. Edn*, **8**, 1 (1969); *Accounts Chem. Research*, **4**, 272 (1971).
10. G. M. Badger, *Aromatic Character and Aromaticity*, Cambridge, 1969.
11. M. G. Evans, *Trans. Farday Soc.*, **35**, 824 (1939).

† In a later development of the theory, it is suggested that radical $(4n + 2)\pi$ Hückel systems are aromatic [M. J. S. Dewar and S. Kirschner, *J. Amer. Chem. Soc.*, **93**, 4290, 4291, 4292 (1971)].

3 Electrocyclic reactions

An electrocyclic reaction is the formation of a σ bond between the termini of a fully conjugated linear π system, or the reverse process. It is therefore a type of intramolecular cycloaddition or retroaddition.

Electrocyclic reactions can be brought about by heat, by ultraviolet irradiation, and sometimes by the use of metal catalysts. They are nearly always stereospecific. In many cases, detection of their stereospecificity depends on distinguishing chemically similar stereoisomers, a problem which has been overcome mainly by the development of spectroscopic methods of structure determination, especially n.m.r. Thus, the recognition that stereospecific electrocyclic reactions form a coherent group extends only over the last few years; even so, the group includes some important synthetic reactions as well as some of the most clearcut examples of the successful predictive power of the orbital symmetry theory.

3.1. Thermal electrocyclic reactions. There are four stereochemically distinguishable ways in which an electrocyclic reaction can take place; two are disrotatory and two conrotatory. These are illustrated in fig. 3.1 for a ring opening.

Not all these modes of ring opening will be distinguishable in a particular case. If A, B, C and D are identical, for example, there is only one possible product. If A, B, C and D are all different groups, there are four possible products. Basically, the theory distinguishes only the disrotatory modes of ring opening from the conrotatory modes, and does not distinguish between the two possible disrotatory modes or conrotatory modes. Predictions as to which of these will be preferred can usually be made on the basis of steric effects, as subsequent examples will show.

The mode of electrocyclic ring opening or closure depends simply on the number of π electrons in the open polyene. A conjugated linear

49

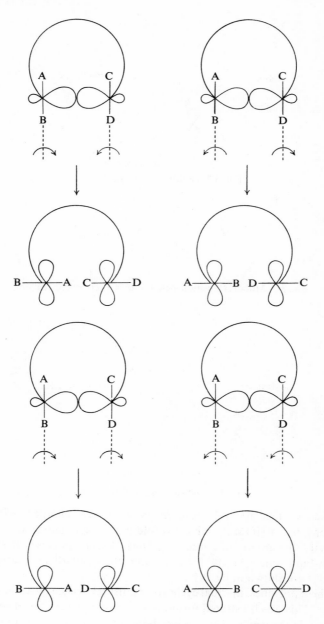

Fig. 3.1

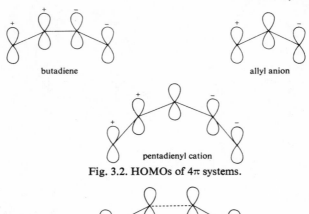

Fig. 3.2. HOMOs of 4π systems.

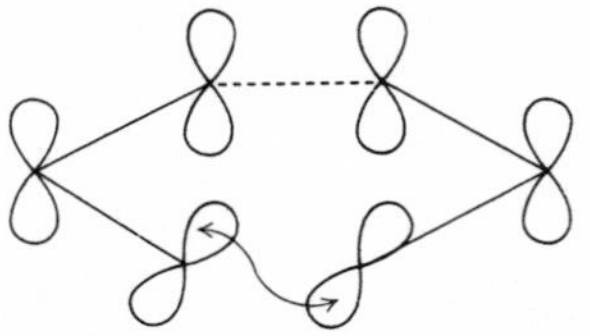

Fig. 3.3. Möbius transition state for $4n$ π electrocyclic reactions.

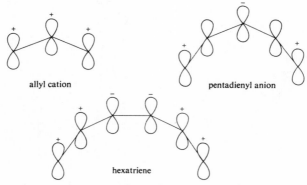

Fig. 3.4. HOMOs of $(4n + 2)$ π systems.

polyene, whether neutral, positive, or negatively charged, with $4n$ π electrons, has a HOMO with a twofold axis of symmetry, in which the terminal lobes are out of phase. Examples of such systems (where $n = 1$) are *cis*-butadiene, the allyl anion, and the pentadienyl cation; the HOMOs are shown in fig. 3.2.

In these polyenes, thermal cyclisation must take place by a conrotatory closure. This was illustrated for butadiene in chapter 2; it is most simply pictured, in the general case, as involving a rotation of the termini of the

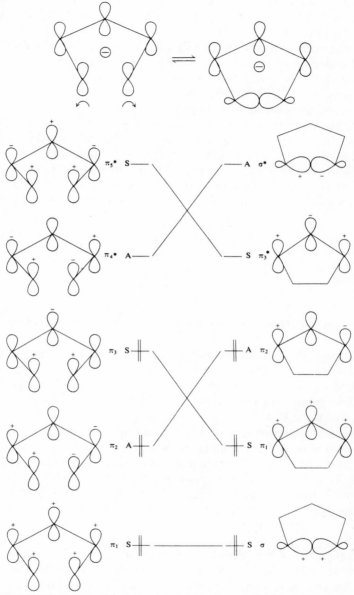

Fig. 3.5. Correlation diagram for disrotatory closure of pentadienyl anion.

HOMO in a direction necessary for overlap to form the new σ bond; or using the aromatic transition state approach, as involving a Möbius type transition state (fig. 3.3).

For linear π systems with $(4n + 2)$ π electrons, the thermal cyclisation is predicted to be disrotatory. Such systems include the allyl cation (two π electrons, $n = 0$), hexatrienes, and the pentadienyl anion (six π electrons, $n = 1$). The HOMOs of these systems have a plane of symmetry, and the terminal lobes in phase (fig. 3.4).

The disrotatory mode of closure can be predicted from the appropriate correlation diagrams, just as for the conrotatory closure of 4π systems. Consider, as an example, the thermal interconversion of the pentadienyl and cyclopentenyl anions. In the disrotatory closure a plane of symmetry is maintained throughout the reaction; the orbital correlation diagram is shown in fig. 3.5. It is a useful exercise to show that for

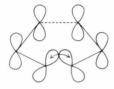

Fig. 3.6. Hückel transition state for $(4n + 2)$ π electrocyclic reactions.

this system, the thermal conrotatory mode is forbidden. (The orbitals involved are the same but their symmetries are different because a twofold axis is maintained throughout, not a plane of symmetry. Thus, $π_1$ of the pentadienyl anion is antisymmetric with respect to the twofold axis, and so on.)

Again, the selection rules can be deduced without recourse to correlation diagrams by concentrating either on the symmetry of the highest occupied orbital, or on the 'aromatic' nature of the transition state. Thus, for a $(4n + 2)$ π system, the in-phase termini of the HOMO must move in a disrotatory fashion to overlap; or, as a Hückel $(4n + 2)$ π system, the aromatic transition state is achieved without sign inversion in the system of p-orbitals (fig. 3.6).

Examples of thermal electrocyclic reactions.
1. *2π Systems.* The cyclopropyl to allyl cation isomerisation (3.1) is a 2π electrocyclic ring opening. The reaction is an important one, especially in bicyclic systems, where it leads to ring expansion (3.2).

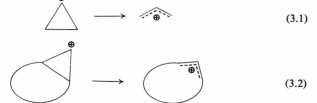

$$(3.1)$$

$$(3.2)$$

It is unlikely that free cyclopropyl cations are intermediates in the reported examples of reactions of this type; the removal of the leaving anion is assisted by the concerted breaking of the cyclopropane C—C bond and the generation of the allyl π system. This is elegantly shown by experiments with bicyclic cyclopropanes which reveal that the leaving group has to be *endo* for the smooth ring opening to take place. The effect is especially marked in cyclopropanes fused to five- or six-membered rings. This result was predicted by Woodward and Hoffmann in 1965, and has been confirmed in several systems since then. One example is provided by the isomeric 6-chlorobicyclo[3,1,0]hexanes (**1**) and (**2**); the *endo*-isomer (**1**) is converted into 3-chlorocyclohexene by heating for three hours at 126°C, but the *exo*-isomer (**2**) was found to be unchanged, even in much more vigorous conditions.[1]

(**1**)   (**2**)

126°C/3 h   250°C/4 h (3.3)

                           no reaction

Woodward and Hoffmann explain the preferred removal of the *endo*-anion as follows. The electrocyclic ring opening is a disrotatory one, and only the disrotatory mode which moves the bridgehead hydrogens outward is geometrically feasible in small fused ring systems. In the *endo*-isomer, therefore, the C—C bond breaks in such a way as to increase the electron density behind the leaving group, and to bring about a type of S_N2 displacement (3.4).

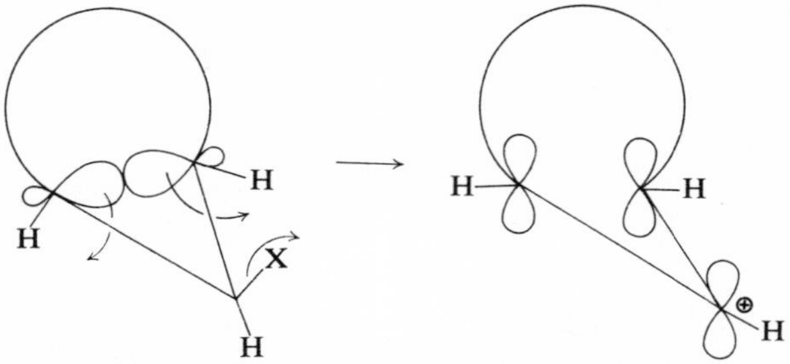

$$(3.4)$$

The effect is less marked in larger fused rings and in monocyclic cyclopropanes, but it still operates. For example, the thermal ring opening of cyclopropane (3) is faster than that of (4) because in (3), the required disrotatory mode allows an outward rotation of the methyl groups, but in (4), the methyls must move towards each other in the transition state.[2]

(3)   (4)

The idea of concerted disrotatory ring opening and displacement of a leaving group is not confined to electrocyclic reactions. Wiberg has observed a similar effect in fused cyclobutane derivatives.[4] An extreme example of the accelerating effect that the concerted reaction can have is provided by the isomers (5) and (6). The rate of solvolysis of the *endo*-isomer (5) is 10^7 times that of (6) and the enthalpy of activation is 12 kcal (50 kJ) mol^{-1} lower. Only in (5) can the disrotatory ring opening and loss of the leaving group be concerted.

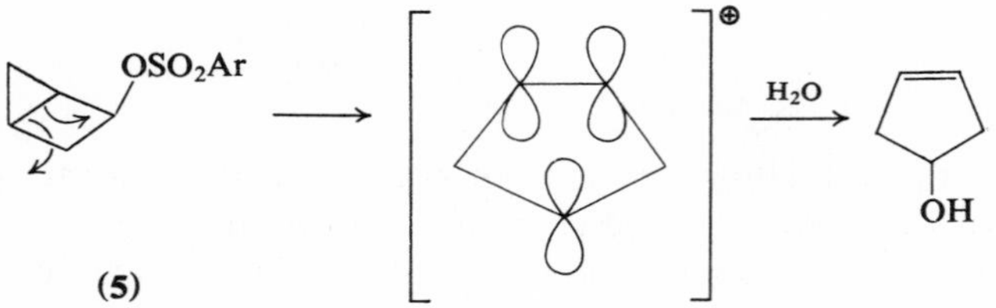

$$(3.5)$$

(5)

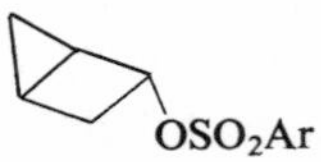

(6)

Other factors which affect the ease of electrocyclic ring opening include the nature of the substituents, which can stabilise or destabilise the developing positive charge, and the relief of strain in small bicyclic systems. Thus, the adjacent oxygen may assist the opening of the ether (7), which goes smoothly at 60°; and the strained bicyclopentane (8) in unstable even at 0°C (the ring opening in this case being facilitated by the relief of ring strain).

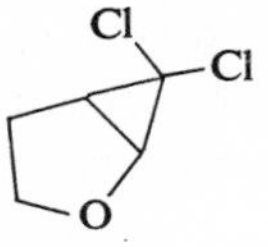

(7) 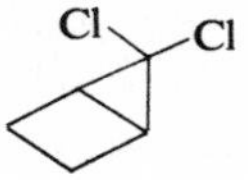   (8)

The cyclopropyl to allyl ring opening is one of the most useful of electrocyclic reactions, especially as a means of ring expansion, since the halogeno-cyclopropanes are readily available by carbene addition to the appropriate olefins.

An interesting extension of the reaction, which involves an iso-electronic heterocyclic ring expansion with the same orbital symmetry requirements, is the thermal conversion of the chloro-aziridine (9) to isoquinoline (3.6).[3]

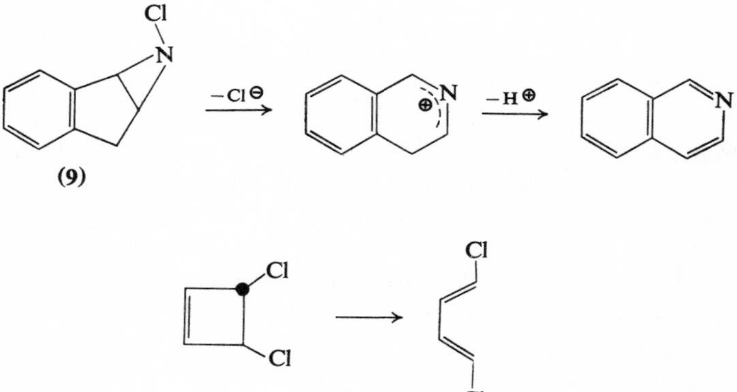

(9)   (3.6)

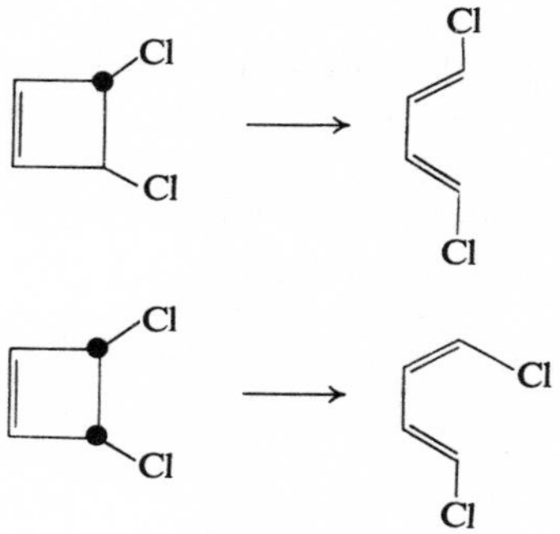

(3.7)

2. *4π Systems.* The thermal ring opening of cyclobutenes to butadienes is stereospecific, and conrotatory, as the theory predicts. In most cases the ring opening goes to completion and there are very few examples of the reverse process, the thermal cyclisation of butadienes. The electrocyclic ring opening takes place smoothly in solution or in the gas phase at temperatures between about 120°C and 200°C, the activation energy being about 35 kcal mol⁻¹ (146 kJ mol⁻¹). Typical examples occur with the *trans-* and *cis-*3,4-dichlorocyclobutenes, which open stereospecifically to *trans,trans-* and *cis,trans-*1,4-dichlorobutadiene, respectively (3.7).[5]

An example of the reverse process is provided by the conrotatory cyclisation of the strained *cis,trans-*1,3-cyclo-octadiene (10) which takes place at 80°C.[6]

(10)   (11)   (3.8)

Fused cyclobutenes are thermally rather stable, especially those in which the second ring is five- or six-membered. For example, the cyclobutene (12) does not isomerise below about 380°C.[5]

(12)

This might seem surprising, since the fusion of a second ring might be expected to increase the strain and make ring opening easier. However, the allowed ring opening is conrotatory, which would require the formation of dienes containing a *trans* fused double bond. The fused cyclobutenes prefer to take a path which leads to the disallowed *cis, cis* cyclic diene. For example, the bicyclo-octene (11) opens at 230–260°C to give *cis,cis-*1,3-cyclo-octadiene (3.9), and not the *cis,trans-*diene (10).[7] The activation energy, 43 kcal mol⁻¹ (180 kJ mol⁻¹), is much higher than for the opening of simple cyclobutenes.

(11)   (3.9)

Conrotatory thermal ring opening is also predicted for reactions of the type (13) → (14), in which the four electrons include a lone pair on the hetero-atom X. Several examples of this type of ring opening are

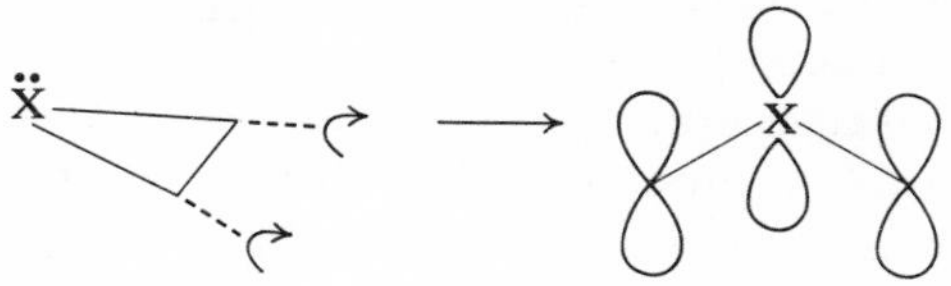

(13)   (14)

known, and where the stereochemistry has been determined, the opening is conrotatory. For example, many aziridines open when heated to about 100°C, and form delocalised 4π intermediates which can sometimes be trapped in cycloaddition reactions. The ring opening is conrotatory, as Huisgen and his co-workers showed for the *cis* and *trans* diesters (15) and (16) (equation 3.10).[8] The stereochemistry was established in the cycloadducts of the 1,3-dipoles with dimethyl acetylene-dicarboxylate.

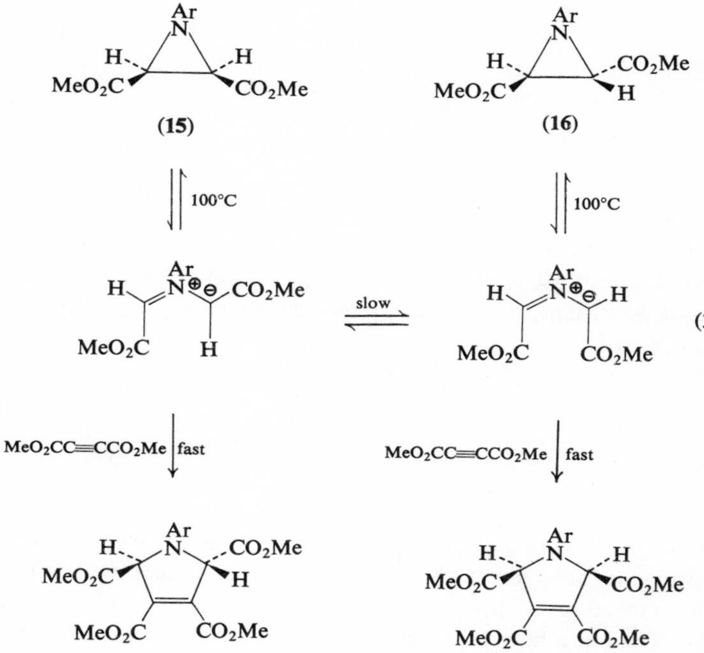

In the absence of a good dienophile, the two open structures are interconvertible, and this provides a mechanism for thermal isomerisation of the aziridines.

Just as with cyclobutenes, fusion of the aziridine in a structure which cannot open by the conrotatory mode inhibits the reaction. An example of such a system is the aziridine (17), which failed to give cycloaddition reactions even at 180°C, presumably because conrotatory opening of the three-membered ring is sterically impossible.[9]

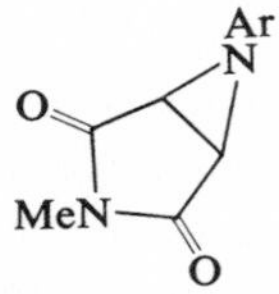

(17)

Another system which shows the same type of thermal ring opening is tetracyanoethylene oxide. The open form has been trapped in cycloaddition reactions: the addition to olefins at 130–150°C is stereospecific (3.11).[10]

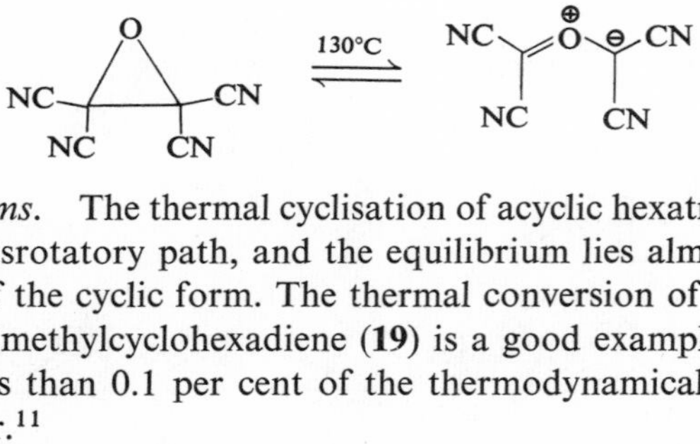

(3.11)

3. 6π *Systems*. The thermal cyclisation of acyclic hexatrienes takes the predicted disrotatory path, and the equilibrium lies almost completely in favour of the cyclic form. The thermal conversion of the triene (18) to the *cis*-dimethylcyclohexadiene (19) is a good example; the product contains less than 0.1 per cent of the thermodynamically more stable *trans* isomer.[11]

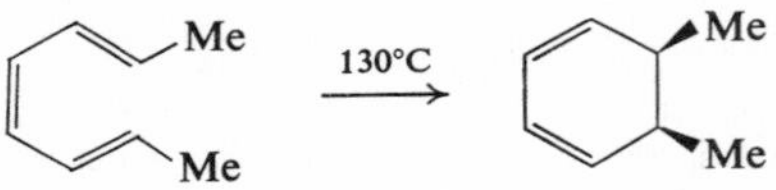

(3.12)

Although a few other acyclic examples of stereospecific isomerisation of hexatrienes are known, especially in natural product work, the most common reactions of this type are in cyclic hexatrienes. Cyclononatriene, cyclo-octatriene and cyclo-octatetraene are systems in which the electrocyclic reaction goes very readily, and they show an interesting trend.[12] The cyclisation of cyclononatriene goes virtually to completion at room temperature, whereas the reverse reaction is slightly more favourable for the cyclo-octatriene equilibrium. For cyclo-octatetraene and bicyclo[4,2,0]octatriene (20), the equilibrium strongly favours cyclo-octatetraene, to such an extent that the half-life of the bicyclo-isomer (20) is only a few minutes at 0°C. Thus, as the bicyclic system becomes more strained, the equilibrium lies more in favour of the open structure.

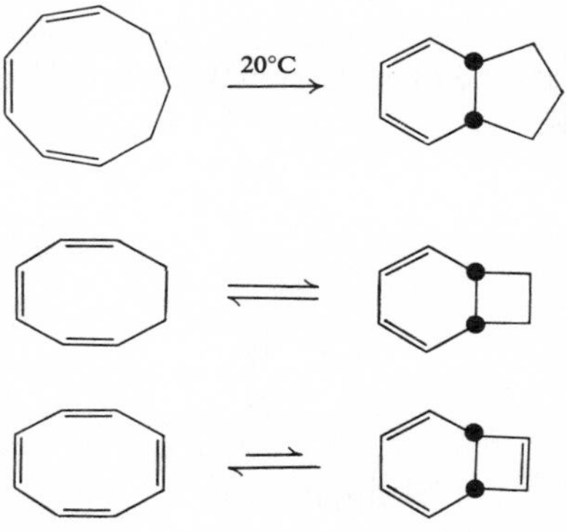

(3.13)

(20)

Norcaradiene and cycloheptatriene structures are similarly inter-related through a disrotatory electrocyclic reaction (3.14). As with any rapidly equilibrating mixture, the structures cannot be distinguished by chemical means – catalytic reduction of a norcaradiene derivative gives a cycloheptane, for example – and the problem of whether a particular compound has the norcaradiene or the cycloheptatriene structure is best solved by the use of high resolution n.m.r.

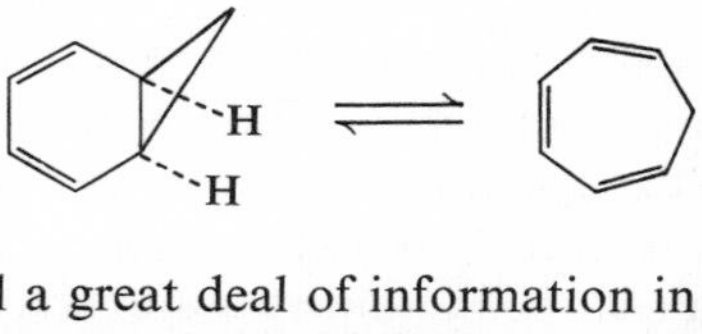

(3.14)

N.m.r. can reveal a great deal of information in structural problems of this type. Not only can it distinguish the valence tautomers, but it can be used to study the rate of interconversion of the tautomers at temperatures where they are in equilibrium.

At temperatures below that required to establish the equilibrium, the n.m.r. spectrum shows only one of the possible isomeric structures, which can therefore be distinguished. Similarly, if the valence tautomers are in rapid equilibrium, but one tautomer is a very minor component (less than about 2 per cent of the total) in the equilibrium mixture, then the minor component is not detected and the n.m.r. spectrum represents only the structure of the major tautomer. These spectra are relatively simple to interpret. An example of the use of n.m.r. to distinguish a non-equilibrating system of this type is provided by the assignment of structure to the first simple norcaradiene to be prepared, the dicyano-derivative (**21**). N.m.r. established the norcaradiene structure rather than the cycloheptatriene one; the signals from the bridgehead hydrogen atoms appear at 6.53 τ.[13]

(**21**)

An even more powerful application of n.m.r. is to equilibrating systems where the tautomers are present in comparable amounts.[14] In these cases the spectra may be temperature dependent. If Δt (s) is the lifetime of a particular tautomer and $\Delta \nu$ (Hz) the difference in chemical shifts of a proton in two different tautomers, then the spectra will be temperature dependent if the product $\Delta t \cdot \Delta \nu$ lies between 10^{-3} and 10. In such systems the spectra normally reflect three structural situations.

1. At low temperatures, the rate of interconversion of the tautomers may be so slow that the spectra of the individual tautomers can be seen, and are relatively invariant within this temperature range.
2. In an intermediate temperature range, the rate of interconversion of the tautomers is comparable with the n.m.r. time scale. The signals due to the individual tautomers broaden, become diffuse and then coalesce to form new signals at positions intermediate between those for the same protons in the separate tautomers.
3. At high temperatures, the tautomers are interconverted too rapidly to be detected individually, and the spectrum is an 'average', not equivalent to any single classical structure. When the sample is cooled, the low temperature spectrum should reappear, showing that no irreversible change has taken place.

Several examples of the use of this technique to detect valence tauto-merism appear in later chapters. An application to the present problem is the interconversion of the two fused norcaradienes (22) and (24), through the valence tautomer (23).[15] In this system the equilibrium strongly favours the norcaradienes, because aromaticity is lost in the cycloheptatriene structure. At room temperature, the spectrum is that of the norcaradiene, with the signal from the *exo*-proton H^b at 8.70 τ and that from the *endo*-proton H^a at 10.35 τ.

$$(22) \qquad\qquad (23) \qquad\qquad (24) \qquad (3.15)$$

When the solution is warmed, the signals become diffuse and slowly merge, until at 180°C they appear as a uniform flat absorption band centred at 9.54 τ, midway between the original signals. Further increase in temperature sharpens the signal but does not alter its position. The other peaks in the spectrum are unchanged, and when the solution is cooled, the original signals reappear.

The explanation is that as the temperature is raised, the two norcar-adienes are interconverted through the cycloheptatriene, though its concentration remains minute. In (24) the *exo*-proton of (22) has become *endo*, and vice versa, so that the signals in the two environments become indistinguishable when the interconversion is rapid.

In most derivatives of this system the equilibrium strongly favours one of the two forms, and the other is not detected. Most simple derivatives exist in the cycloheptatriene form, the dicyano-derivative (21) being a rare exception.[16] In fused derivatives, like (22), the norcaradiene structure may be favoured for special reasons, such as the loss of re-sonance energy in the cycloheptatriene.

Benzene oxide (25) and oxepin (26) are another example of a pair of valence tautomers interconvertible by disrotatory electrocyclic reactions. The system has been extensively studied by Vogel and his colleagues.[17] Benzene oxide and other arene oxides may be intermediates in the oxidative metabolism of aromatic substrates to phenols and catechols. 1,2-Naphthalene oxide (27) has actually been detected as an intermediate in the oxidation of naphthalene in a biological system.[18]

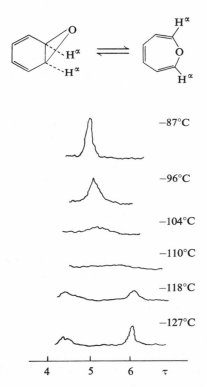

(25)   **(26)**   **(27)**

N.m.r. spectra of benzene oxide–oxepin give a great deal of information about the equilibrium. At room temperature, the tautomers are in equilibrium with appreciable amounts of both isomers in the equilibrium mixture (the exact proportions depend on the solvent and on the temperature). The n.m.r. spectrum is a complex one, showing 'weighted

Fig. 3.7. Signals of α-hydrogens in ¹H n.m.r. spectrum (in CF_3Br/pentane) of benzene oxide – oxepin. Adapted from fig. 572.3 in the article by E. Vogel and H. Günther, *Angewandte Chemie International Edition*, **6**, 390 (1967).

average' signals from the protons in the rapidly interconverting tautomers. Below about −80° the signals broaden and eventually separate
into new signals which can be assigned to protons in the individual
tautomers. The signal for the α hydrogens at low temperatures is shown
in fig. 3.7. The broad peak which appears at about 5 τ in the spectrum
run at −87°C represents an 'average' for the α-hydrogens in the rapidly
interconverting tautomers. As the temperature is lowered the signal
becomes more diffuse, and below −113°C it separates into two new
signals at 4.3 τ and 6.0 τ. These can be assigned to the α-protons in
oxepin and benzene oxide, respectively.

The activation energy barrier to the interconversion is much too low
for the individual tautomers to be isolated at room temperature. As the
temperature is lowered, the proportion of benzene oxide in the equilibrium mixture increases, since the unfavourable entropy term $T\Delta S$
becomes less significant.

Fused systems of this type can exist exclusively in the oxepin form or
exclusively in the benzene oxide form, just like their all-carbon analogues. 1,2-Naphthalene oxide (**27**), for example, has n.m.r. and u.v.
spectra consistent only with this structure, not the oxepin structure. The
spectra resemble those of the corresponding norcaradiene (**22**) very
closely. On the other hand, the oxepin (**28**) exists in the open form because
this contains a nearly planar, aromatic 10π carbon skeleton, whereas
the tautomer (**29**) does not. A similar tautomeric equilibrium might be
expected for the benzene imine–azepine system, (**30**) ⇌ (**31**). However,

(**28**)   (**29**)

(3.16)

(**30**)   (**31**)

(3.17)

(**33**)   (**32**)

the equilibria normally favour the azepine structures very strongly in the systems studied so far; the n.m.r. spectra are not temperature dependent.[19] An exception is the azepine (32), which does show a temperature dependent spectrum;[20] the electron withdrawing substituents may help to stabilise the benzene imine structure (33).[16]

One other example of disrotatory 6 π electron cyclisation may be mentioned. The disrotatory cyclisation of a pentadienyl anion, for which the correlation diagram was constructed earlier, is probably implicated in the base induced conversion of hydrobenzamide (34) into amarine (35), in which the phenyl groups are *cis*.[21]

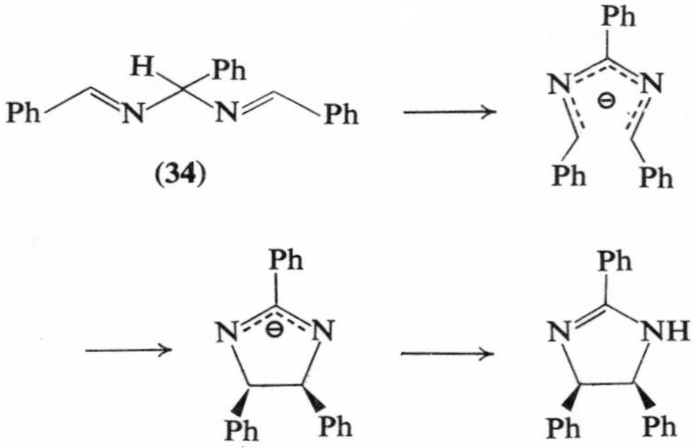

(3.18)

4. *Other systems.* The selection rules apply to 8 π electron and higher systems. Thus, the thermal cyclisation of an 8π system should be conrotatory, for a 10π system it should be disrotatory, and so on. Examples of such reactions are few, but they conform to the expected pattern. One example is the cyclisation of the tetraenes (36) and (37) by conrotatory closure.[22]

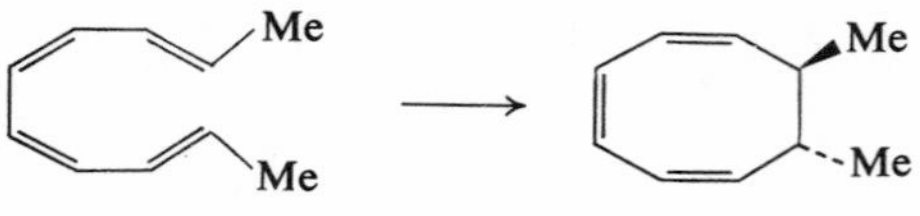

(3.19)

(3.20)

3.2. Photochemical electrocyclic reactions. Electrocyclic reactions brought about by irradiation with ultraviolet light are stereospecific, but in precisely the opposite sense to the thermal processes. If the thermal reaction is disrotatory, then the photochemical reaction is conrotatory, and vice versa. This can be important for synthetic applications, since by choosing the appropriate conditions, the stereochemistry of the product can be selected.

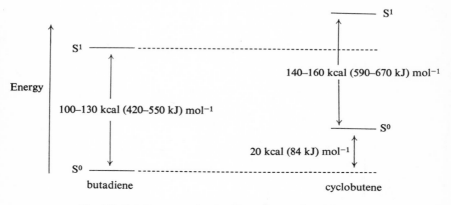

Fig. 3.8. Relative energy levels of butadiene and cyclobutene.

The LVMO of a conjugated linear polyene is of opposite symmetry to that of the HOMO. Thus the simple picture of orbital symmetry control predicts that excitation of an electron into the LVMO will reverse the direction of ring closure or ring opening, compared with that of the ground state system. For 2, 6, ..., $(4n + 2)$ π systems, photo-chemically-induced ring closure should be conrotatory, and for 4, 8, ..., $4n$ π systems, it should be disrotatory.

This simple picture agrees remarkably well with the observed reaction paths. However, it does raise problems, which have been discussed by Oosterhoff and others.[23] Consider the butadiene–cyclobutene ring closure, which can be brought about photochemically and which in substituted butadienes has been shown to be disrotatory. The energy levels of the ground states and the spectroscopic singlet states of butadiene and cyclobutene are shown in fig. 3.8; the diagram indicates that the spectroscopic singlet state of cyclobutene is about 50–60 kcal mol^{-1} (210–250 kJ mol^{-1}) higher in energy than that of butadiene. Simple orbital correlation diagrams predict a correlation between these two states in the 'photochemically allowed' disrotatory ring closure, but the energy difference between the excited states makes this extremely unlikely.

Van der Lugt and Oosterhoff have carried out valence bond calculations which provide a possible solution to the problem, but which indicate that the photochemical process is more complicated than the

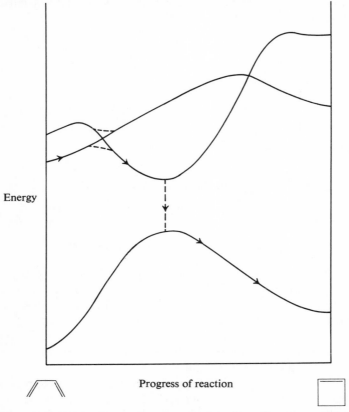

Fig. 3.9. Photochemical disrotatory conversion of butadiene into cyclobutene.

simple theory suggests. As the disrotatory closure of butadiene to cyclobutene proceeds, the first excited singlet increases in energy. However, a higher excited singlet state *decreases* in energy and goes through a minimum before increasing in energy again. This energy minimum is at about the same point along the reaction path as the energy maximum in the ground state closure. The proposal is that nuclear vibrations cause the molecule to pass adiabatically into the second excited state at the point along the path where the excited states coincide, and then it drops down from the energy minimum to the ground state contour, thus going on to give ground state cyclobutene (fig. 3.9).

For the conrotatory closure there is no comparable energy minimum in the excited state, so the butadiene chooses the disrotatory path. The direction of closure is still as the simple theory predicts, therefore, but the cyclobutene is formed in the electronic ground state.

Photochemical butadiene–cyclobutene interconversions are disrotatory, and most involve conversions of butadienes into cyclobutenes, not the reverse. This is simply a result of the fact that the diene is a stronger absorber of light at the wavelengths used in the reactions, so it is the diene which is excited. The most instructive examples are those where the diene forms part of a ring so that a fused cyclobutene is formed (3.21).

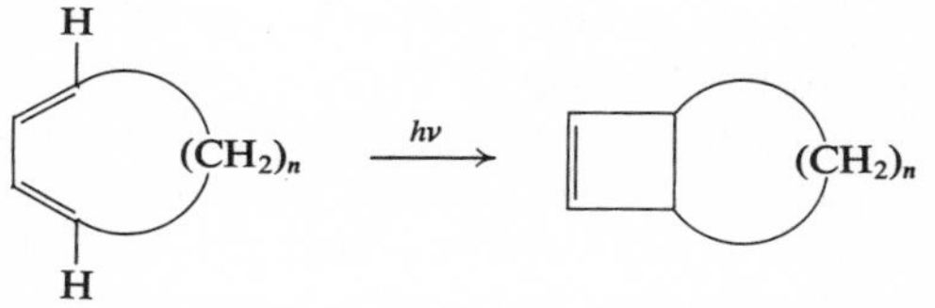

$$(3.21)$$

Since the reverse reaction is thermally disallowed, the products may be unexpectedly stable. Thus, for example, cyclopentadiene can be partially converted photochemically into bicyclo[2,1,0]pentene, which despite its strained structure has a half-life of about two hours at room temperature (3.22).[24]

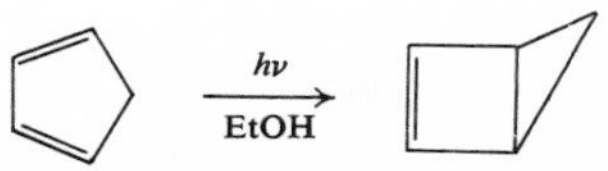

$$(3.22)$$

The photochemical equilibration of benzene derivatives and Dewar benzenes (38) can be regarded as an electrocyclic reaction of this type (3.23). The surprising thermal stability of Dewar benzenes can again be explained by the fact that their isomerisation to benzenes, which

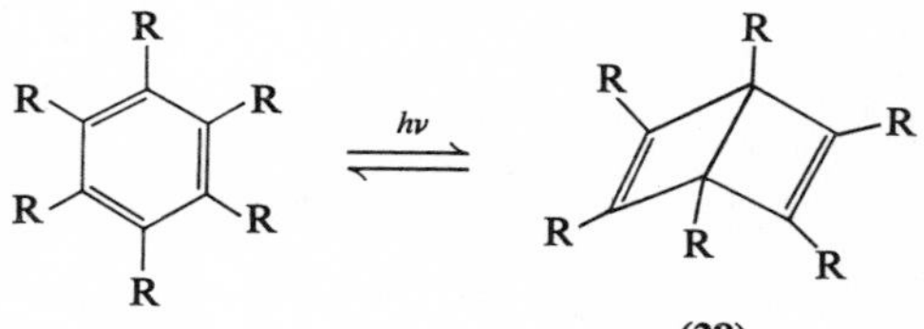

$$(3.23)$$

(38)

must be disrotatory because of geometrical constraints, is a thermally disallowed reaction. 2-Pyrone (**39**) undergoes a similar photochemical equilibration, and in the presence of iron pentacarbonyl, the bicyclic tautomer is decarboxylated to give cyclobutadiene-iron tricarbonyl complex (3.24). This is a useful synthetic application of the photochemical ring closure.[25]

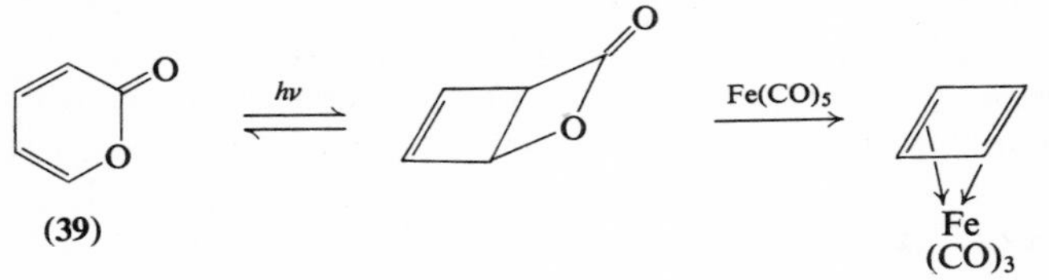

(3.24)

(**39**)

For 6π systems, the photochemical cyclisation is conrotatory, again in striking contrast to the thermal reactions. Examples of photochemical hexatriene cyclisations in natural products (chapter 2) were a stimulus for the development of the theory of orbital symmetry control. Other examples involve isomerisations of medium and large ring polyenes. The all *cis* cyclodecapentaene (**40**), for example, equilibrates photochemically at low temperatures with *trans*-9,10-dihydronaphthalene, by a conrotatory six-electron electrocyclic reaction, but it is converted thermally into *cis*-9,10-dihydronaphthalene by disrotatory closure.[26]

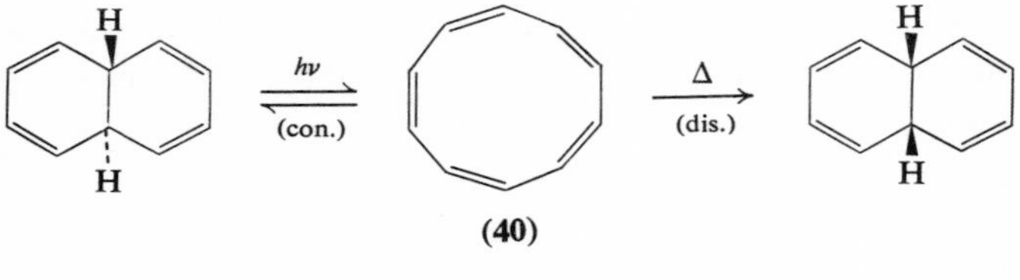

(3.25)

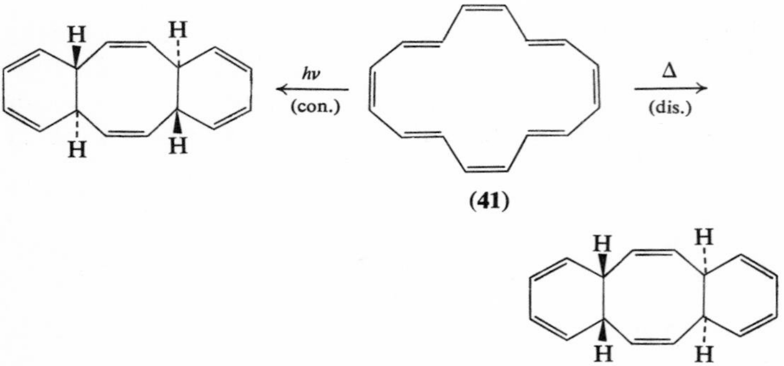

(**40**)

(**41**)

(3.26)

There is similar stereospecificity in the thermal and photochemical electrocyclic reactions of [16]annulene (**41**) (equation 3.26); these involve double disrotatory and double conrotatory ring closures, respectively.[27]

3.3. Metal catalysed electrocyclic reactions. Silver ions and other metal catalysts can have a dramatic effect on the rate of ring opening of strained cyclobutene derivatives.[28] These reactions are necessarily disrotatory because of the stereochemistry of the systems, and as such are thermally disallowed as concerted processes. The slow thermal conversion of bicyclo[4,2,0]octene (**11**) to *cis,cis*-1,3-cyclo-octadiene has already been referred to; the reaction is enormously speeded up in the presence of silver ions. Similarly, the disrotatory opening of the fused cyclobutenes (**42**) and (**43**) is catalysed by silver ions (3.27, 3.28). The conversion of (**42**) to dibenzocyclo-octatetraene is virtually instantaneous at room temperature in the presence of silver ions, and the activation energy is reduced from 23 kcal mol^{-1} (96 kJ mol^{-1}) for the uncatalysed reaction to about 8 kcal mol^{-1} (33 kJ mol^{-1}) for the catalysed process.

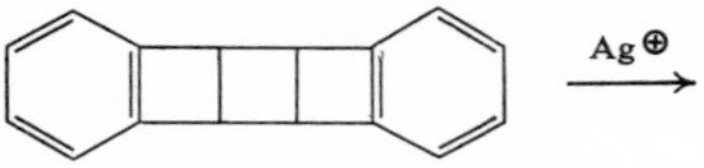

(**42**)

(not isolated)   (3.27)

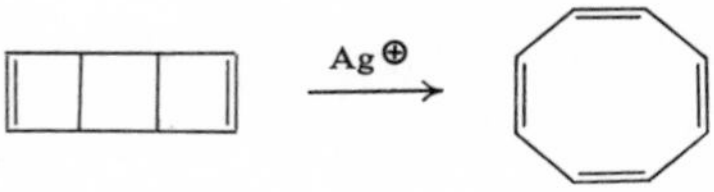

(3.28)

(**43**)

The thermal conversion of hexamethyl Dewar benzene (**44**) into hexamethylbenzene can also be catalysed; in the presence of a rhodium catalyst the activation energy is lowered by about 12 kcal mol^{-1} (50 kJ mol^{-1}).

There have been several tentative suggestions of ways in which the catalysts might operate. First, it is possible that in the presence of the

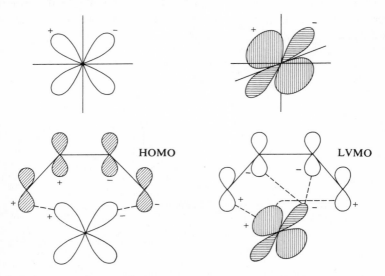

(3.29)

(44)

metal, a completely different mechanism operates, which is a route of low energy and involving one or more intermediates. This seems to be so, for example, in the case of certain catalysed sigmatropic reactions (chapter 7), in which metal hydride intermediates are probably involved, and in some cycloadditions. An attractive alternative suggestion is that the metal supplies filled and empty orbitals of the right symmetry to interact with the LVMO and HOMO of the diene, thus effectively transferring electron density from the HOMO to the LVMO (fig. 3.10) and therefore reversing the symmetry requirements of the system; that is, making the forbidden disrotatory reaction an allowed one.

HOMO

LVMO

Fig. 3.10. (Top) Orthogonal vacant and filled metal d-orbitals; (below) Interaction of metal d-orbitals with diene.

Unfortunately, this explanation cannot simply be applied to silver ions and other d^{10}-systems which have no suitable vacant orbital for interaction.

Another explanation has been expounded by van der Lugt;[29] namely that the reaction remains forbidden in the presence of the metal, but that the activation energy is lowered because the excited electronic configurations of the metal–substrate complex are much lower in energy than those of the substrate alone. This principle can also be applied to some of the catalysed cyclodimerisations of olefins (chapter 6).

At the time of writing, the mechanisms by which these catalysts can operate are still the subject of debate in the literature.[30] It seems unlikely that one single type of mechanism will suffice to explain all the observed cases of catalysis, however.

REFERENCES

1. M. S. Baird, D. G. Lindsay, and C. B. Reese, *J. Chem. Soc.* (C), 1969, 1173.
2. M. S. Baird and C. B. Reese, *Tetrahedron Letters*, 1969, 2117.
3. D. C. Horwell and C. W. Rees, *Chem. Comm.*, 1969, 1428.
4. K. B. Wiberg, V. Z. Williams, and L. E. Friedrich, *J. Amer. Chem. Soc.*, **92**, 564 (1970).
5. R. Criegee, D. Seebach, R. E. Winter, B. Börretzen, and H.-A. Brune, *Chem. Berichte*, **98**, 2339 (1965).
6. K. M. Shumate, P. N. Neuman, and G. J. Fonken, *J. Amer. Chem. Soc.*, **87**, 3996 (1965).
7. G. R. Branton, H. M. Frey, and R. F. Skinner, *Trans. Faraday Soc.*, **62**, 1546 (1966).
8. R. Huisgen, W. Scheer, and H. Huber, *J. Amer. Chem. Soc.*, **89**, 1753 (1967).
9. R. Huisgen and H. Mäder, *Angew. Chem. Int. Edn*, **8**, 604 (1969).
10. W. J. Linn and R. E. Benson, *J. Amer. Chem. Soc.*, **87**, 3657 (1965).
11. E. N. Marvell, G. Caple, and B. Schatz, *Tetrahedron Letters*, 1965, 385.
12. D. S. Glass, J. W. H. Watthey, and S. Winstein, *Tetrahedron Letters*, 1965, 377.
13. E. Ciganek, *J. Amer. Chem. Soc.*, **89**, 1454 (1967).
14. G. Schröder, *Angew. Chem. Int. Edn*, **4**, 752 (1965).
15. E. Vogel, D. Wendisch, and W. R. Roth, *Angew. Chem. Int. Edn*, **3**, 443 (1964).
16. A simple explanation, based on frontier orbital theory, has been given for the effect of substituents on the equilibrium: R. Hoffmann, *Tetrahedron Letters*, 1970, 2907.
17. Review: E. Vogel and H. Günther, *Angew. Chem. Int. Edn*, **6**, 385 (1967).
18. D. M. Jerina, H. Ziffer, and J. W. Daly, *J. Amer. Chem. Soc.*, **92**, 1056 (1970).
19. L. A. Paquette, J. H. Barrett, and D. E. Kuhla, *J. Amer. Chem. Soc.*, **91**, 3616 (1969).
20. H. Prinzbach, D. Stusche, and R. Kitzing, *Angew. Chem. Int. Edn*, **9**, 377 (1970).
21. D. H. Hunter and S. K. Sim, *J. Amer. Chem. Soc.*, **91**, 6203 (1969).
22. R. Huisgen, A. Dahmen, and H. Huber, *J. Amer. Chem. Soc.*, **89**, 7130 (1967).
23. W. Th. A. M. van der Lugt and L. J. Oosterhoff, *J. Amer. Chem. Soc.*, **91**, 6042 (1969).
24. J. I. Brauman, L. E. Ellis, and E. E. van Tamelen, *J. Amer. Chem. Soc.*, **88**, 846 (1966,

25. E. J. Corey and J. Streith, *J. Amer. Chem. Soc.*, **86**, 950 (1964).
26. S. Masamune and R. T. Seidner, *Chem. Comm.*, 1969, 542.
27. G. Schröder, W. Martin, and J. F. M. Oth, *Angew. Chem. Int. Edn*, **6**, 870 (1967).
28. Reviews: R. Pettit, H. Sugahara, J. Wristers, and W. Merk, *Discussions Faraday Soc.*, **47**, 71, (1969); F. D. Mango, *Advances in Catalysis*, **20**, 291 (1969).
29. W. Th. A. M. van der Lugt, *Tetrahedron Letters*, 1970, 2281.
30. J. Wristers, L. Brener, and R. Pettit, *J. Amer. Chem. Soc.*, **92**, 7499 (1970); F. D. Mango, *Tetrahedron Letters*, 1971, 505.

4 Cycloadditions: Introduction

Cycloadditions form a very extensive and rapidly expanding area of chemistry, the synthetic potential and theoretical significance of which have only fairly recently been fully recognised.

A cycloaddition is a process in which two or more reactants combine to form a stable cyclic molecule during which no small fragments are eliminated and σ bonds are formed but not broken.[1,2] This definition covers reactions like the Diels–Alder reaction (4.1a) and the dimerisation of olefins (4.1b) but excludes such processes as the Dieckmann cyclisation (4.1c). The majority of cycloadditions involve the formation of two new σ bonds, as in the Diels–Alder reaction, but electrocyclic reactions like the cyclisation of butadiene in which only one new σ bond is formed (4.1d) (together with a π bond) can also be considered as intramolecular cycloadditions. In cyclotrimerisation three new σ bonds are formed.

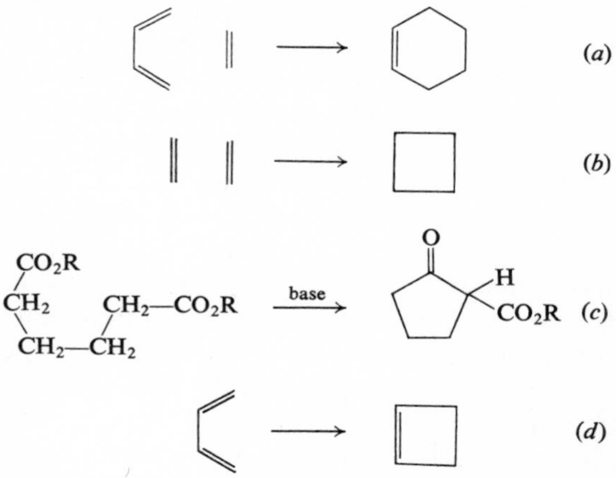

Since cycloaddition is such a varied process further classification is often desirable. Huisgen[2] has suggested that such a classification should be independent of mechanism and his system uses only the number of ring atoms provided by each component. Thus the Diels–Alder reaction (4.1*a*) is a 4 + 2 cycloaddition, and the olefin dimerisation (4.1*b*) a 2 + 2 cycloaddition. Another system recognises the fundamental importance of the number of electrons involved in the cycloaddition and classifies the reactions on this basis. The Diels–Alder reaction thus becomes a $(4\pi + 2\pi)$ process and the olefin dimerisation a $(2\pi + 2\pi)$ one.

In these two examples, the number of electrons happens to be the same as the number of atoms in each component. This is not always the case; in 1,3-dipolar cycloaddition, for example, the 1,3-dipole has 4π electrons distributed over three atoms. The addition of a 1,3-dipole to an olefin is therefore a 3 + 2 addition according to the first classification but $(4\pi + 2\pi)$ according to the second (4.2).

$$\text{(4.2)}$$

4.1. Selection rules for thermal polyene cycloadditions. Many cycloadditions show all the characteristics of concerted processes but others are obviously stepwise and involve an intermediate which is either a zwitterion or a diradical. Why some cycloadditions should be concerted and others not has intrigued and stimulated chemists for many years. We are now in a position to rationalise these differences along the lines described in chapter 2.

Perhaps the simplest approach uses frontier orbital theory. In the concerted cycloaddition of two polyenes, bond formation at each terminus must be developed to some extent in the transition state. Thus orbital overlap must occur simultaneously at both termini. For a low energy concerted process (an allowed reaction) to be possible such simultaneous overlap must be geometrically feasible and must also be potentially bonding.

Suprafacial, suprafacial addition. Suprafacial, suprafacial (s,s) approach of two polyenes (fig. 4.1) is normally sterically suitable for efficient orbital overlap. The vast majority of concerted additions involve the s,s approach.

However, this type of overlap will only be energetically favourable when the HOMO of one component and the LVMO of the other can interact in a bonding fashion at both termini. Thus these orbitals must be of the correct phase or symmetry. In the Diels–Alder reaction of a diene with a monoene the HOMO and the LVMO of each reactant are of the appropriate symmetry so that mixing of these orbitals will result in simultaneous potential bonding character between the terminal atoms (fig. 4.2).

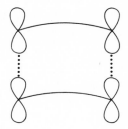

Fig. 4.1. Suprafacial, suprafacial approach of the two polyenes.

In contrast, similar concerted s,s approach of two olefins does not lead to a stabilising interaction since the HOMO and LVMO are of incompatible phase for simultaneous bonding interaction to occur at both termini (fig. 4.3). Thus the initial approach of reactants for a concerted s,s addition is favourable for the Diels–Alder reaction, which is therefore an allowed process, but not for olefin dimerisation, which is therefore disallowed.

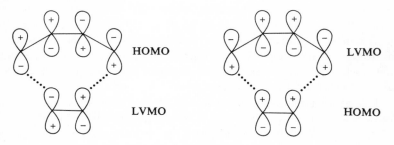

Fig. 4.2

The symmetries of the HOMO and LVMO of polyenes depend on the number of π electrons in the polyene. Thus the favourability of initial interaction for concerted addition also depends on the number of π electrons provided by each component. The total number of π electrons is therefore fundamental to whether a concerted s,s cycloaddition is allowed. In general, s,s cycloaddition can be concerted for $(4n + 2)$ π electrons but not for $4n$. Exactly the same conclusion is reached using the aromatic transition state approach. Only those cycloadditions which have a 'quasi planar' Hückel aromatic $(4n + 2)$ π electron transition state can be concerted (§ 2.4).

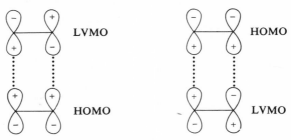

Fig. 4.3

So far with the frontier molecular orbital approach, we have considered only whether the *initial* interaction between the reactants is bonding or antibonding in order to decide whether a concerted cycloaddition is possible. It has been assumed that if this initial interaction is favourable the trend will continue throughout the process and lead to a low activation energy. Correlation diagrams enable the interaction of the reactant orbitals to be followed right through to product orbitals. For allowed, concerted processes reactant bonding orbitals must be able to be transformed into product bonding orbitals without any change of symmetry. Correlation diagrams lead to exactly the same conclusions as frontier orbital theory. This is illustrated for a typical allowed reaction, the Diels–Alder reaction, and for a typical disallowed reaction, the $_\pi 2_s + _\pi 2_s$ dimerisation of ethylene.

The correlation diagram for the $_\pi 4_s + _\pi 2_s$ Diels–Alder reaction is shown in fig. 4.4. The symmetry of the orbitals is labelled with respect to the plane bisecting both reactants. The bonding reactant orbitals are all transformed into bonding product orbitals without a change in symmetry. There is thus no unfavourable rise in energy as the reaction proceeds and a concerted transition is favoured. Alternative stepwise mechanisms would involve diradical or zwitterion formation and so, except in cases where such intermediates are particularly stabilised, these are unlikely to compete with the concerted process.

Note that the monoene π orbital correlates with the cyclohexene π orbital. This is because levels of like symmetry cannot cross.

A similar correlation diagram for the $_\pi 2_s + {}_\pi 2_s$ dimerisation of olefins is shown in fig. 4.5. Here, the orbitals are classed as symmetric or antisymmetric with respect to a plane through the π systems, just as in fig. 4.4. The diagram shows the typical characteristic of a disallowed process – the correlation of a bonding reactant orbital with an antibonding product orbital.

A more sophisticated version of this correlation diagram is shown in fig. 4.6. The degeneracy of the reactant orbitals is removed by their classification with respect to the two mirror planes m_1 and m_2. The SA combination increases in energy as the transition state is approached and the electrons from this orbital should ultimately enter the bonding AS cyclobutane orbital because a crossing of energy levels occurs at high energy. However, this energy barrier is sufficiently high to make it likely that alternative stepwise processes are more favourable.

Other geometries of approach. Two polyenes can also overlap in the suprafacial, antarafacial (s,a) sense as shown in fig. 4.7.

This has different stereochemical consequences: addition to one component is *cis*, to the other *trans*. The efficiency of this type of orbital overlap will depend greatly on the geometry of the reactants, but it is normally considerably less than for the s,s mode. This s,a approach of two olefin molecules leads to a bonding interaction (*a* of fig. 4.8) but similar approach of a diene and a monoene, regardless of which is the antarafacial or suprafacial component, gives an initial antibonding interaction (*b*). The former is therefore allowed, although geometrically improbable, and the latter disallowed.

In general, concerted s,a addition is allowed when the total number of π electrons is $4n$ but disallowed for $(4n + 2)$ π electrons; that is, a concerted cycloaddition involving $4n$ electrons must have one antarafacial component. Such reactions will involve twisted, Möbius type aromatic transition states (§ 2.4).

Antarafacial, antarafacial (a,a) addition (*trans* to each component) would be allowed for the reaction of a monoene with a diene (fig. 4.9) and for any $(4n + 2)$ π electron cycloaddition, but disallowed for olefin–olefin or any other $4n$ π electron cycloaddition. It is important to emphasise again, however, that some allowed modes of concerted addition such as these may be very unfavourable or impossible because for steric reasons the reactant orbitals cannot overlap effectively.

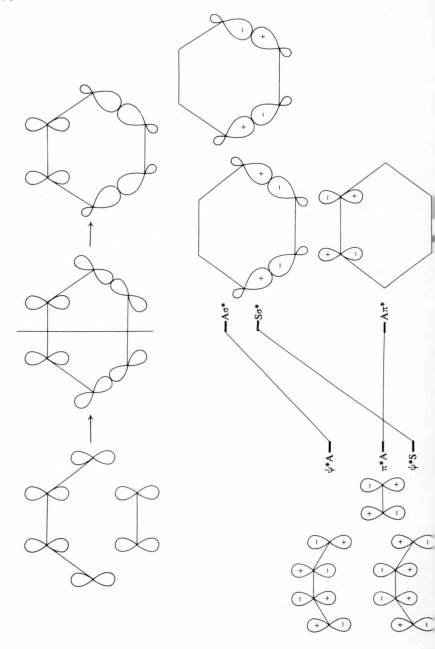

$A\sigma^*$

$S\sigma^*$

ψ^*A

$A\pi^*$

π^*A

ψ^*S

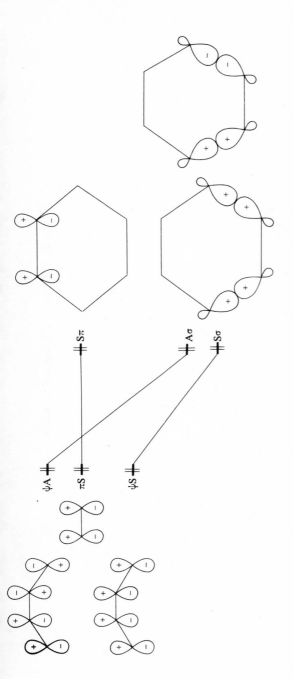

Fig. 4.4. Correlation diagram for the Diels–Alder reaction.

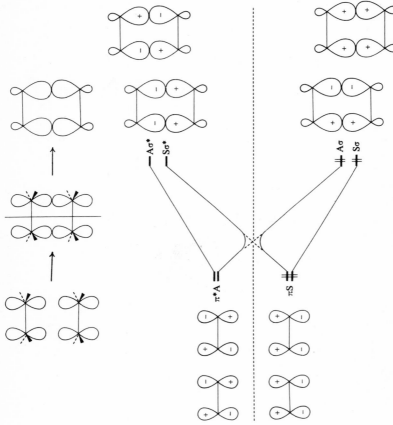

Fig. 4.5. Correlation diagram for the dimerisation of ethylene.

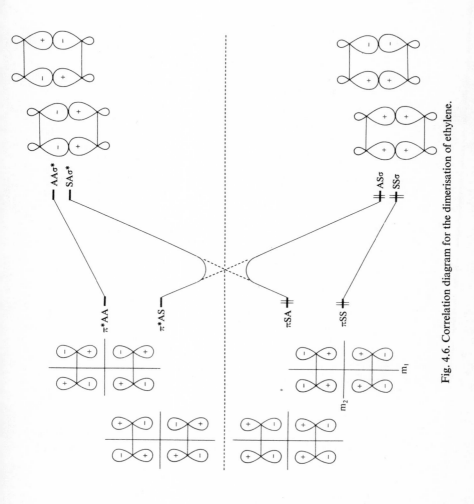

Fig. 4.6. Correlation diagram for the dimerisation of ethylene.

81

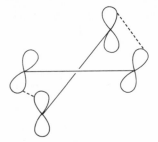

Fig. 4.7. Suprafacial, antarafacial approach of two polyenes.

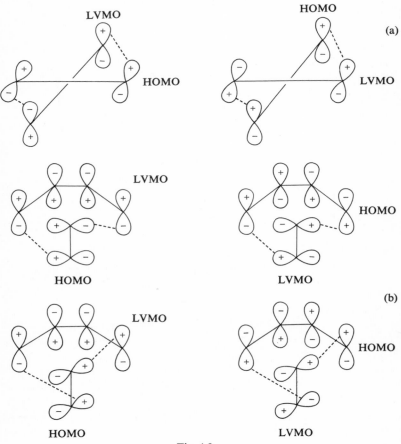

Fig. 4.8

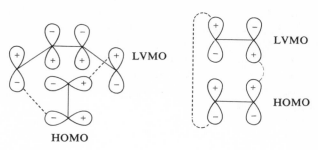

Fig. 4.9

The general selection rules for thermal cycloadditions of two components are summarised as follows:

Concerted s,s or a,a addition is allowed for a total of $(4n + 2)\,\pi$ electrons.
Concerted s,a addition is allowed for a total of $4n\,\pi$ electrons.

These rules are based on the symmetry of the orbitals of acyclic polyenes. When the polyenes form part of cyclic systems their orbital symmetry characteristics are generally unchanged. Simple substituents and even hetero-atoms in the polyene system may alter the energies of the orbitals involved, but probably not their symmetries. The rules are therefore widely applicable. However, if there is any doubt that the symmetries of the reactant orbitals can be inferred from those of the simple open-chain polyene the symmetries should be checked by calculation.

It can be seen that a concerted mode of addition is always formally possible, if not by an s,s route, then by an a,s route. However, not all cycloadditions are concerted because, as already pointed out, the geometry necessary for efficient orbital overlap in the allowed mode may be impossible to attain for steric reasons. Thus, for example, concerted $_{\pi}2_s + {}_{\pi}2_a$ cycloaddition of two olefins is geometrically unfavourable, and since the geometrically favourable $_{\pi}2_s + {}_{\pi}2_s$ addition is forbidden as a concerted process, $2 + 2$ additions are almost always stepwise. On the other hand, the allowed $_{\pi}4_s + {}_{\pi}2_s$ Diels–Alder reaction involves geometrically favourable orbital overlap, and Diels–Alder reactions are mostly concerted and almost invariably occur through this sterically most favourable mode rather than via the allowed but geometrically unfavourable $_{\pi}4_a + {}_{\pi}2_a$ mode. Even when the concerted pathway is both allowed and geometrically favourable, substituents in the reactants can so stabilise an intermediate that an alternative stepwise route may be competitive, and occasionally, predominant.

In some cycloadditions more than one relative orientation of reactants is possible for a particular mode of addition. For example N-substituted azepines undergo allowed $_\pi 4_s + _\pi 6_s$ dimerisation through the transition state (1) rather than (2).

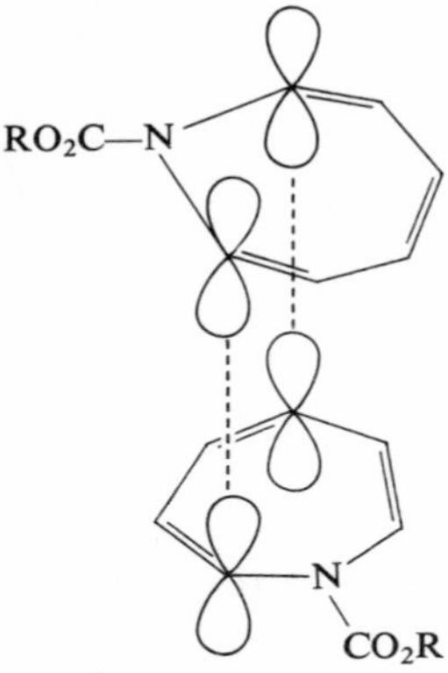

(1) 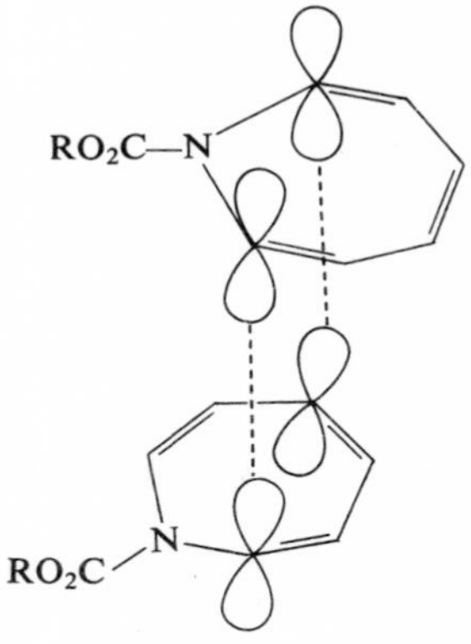   (2)

This can be explained by secondary orbital interactions favouring or disfavouring one particular transition state relative to the other (see chapters 5 and 6). These effects are discussed in more detail with the aid of specific examples in §§ 5.1 and 6.9.

Cycloadditions with more than two components. The above types of argument apply to cycloadditions of more than two components but the geometrical considerations are more complicated. For a $2\pi + 2\pi + 2\pi$ concerted cycloaddition the s,s,s and the three modes of a,a,s (a,a,s; a,s,a; s,a,a) are allowed but the a,a,a and three modes of a,s,s are not. However, concerted intermolecular cycloadditions of more than two components are most unlikely to occur because of the unlikelihood of three-body collisions. Those that are known have at least two of the components constrained in one molecule (§ 5.6).

4.2. Other types of cycloaddition. According to frontier orbital theory, the overriding factor in determining the selection rules for polyene cycloadditions is the symmetry of the HOMO and LVMO of the reactants; this in turn is related to the number of electrons involved in each of the reactants. Any component other than a neutral polyene which can supply the same number of electrons in an orbital of the same symmetry, is potentially able to participate in a cycloaddition in place of the polyene. Some of the more important reactions of this type are briefly outlined here; they are discussed in more detail in chapters 5 and 6.

Three atom components. Cycloadditions involving allylic systems. An allyl cation has two π electrons and a HOMO of the symmetry shown in fig. 4.10(*a*). It is thus equivalent to an olefin in terms of orbital symmetry and should be capable of participating in cycloadditions as the two electron component. Such reactions are known (§ 5.5).

Three atom components with four electrons and the symmetry of the allyl anion, as shown in fig. 4.10(*b*), should similarly be capable of the same types of cycloadditions as dienes. These reactions are well known; they are commonly referred to as 1,3-*dipolar cycloadditions* (§ 5.3).

(*a*)   (*b*)

Fig. 4.10. HOMO of (*a*) allyl cation; (*b*) allyl anion.

One atom components. Cheletropic reactions. Some small molecules have a filled and vacant orbital available on the same atom for bonding to other atoms. Sulphur dioxide, carbon monoxide, and singlet carbenes are examples: in each case one atom (sulphur or carbon) has a lone pair of electrons in the plane of the molecule and a vacant p-orbital orthogonal to it.

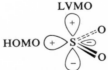

In cycloadditions and eliminations involving such molecules, the σ bonds which are formed or broken are to the same atom; for example, the addition and elimination of sulphur dioxide shown in equation 4.3. Woodward has proposed that these should be called *cheletropic reactions*[3] (Greek, chele – claw).

$$\text{\includegraphics[width=2cm]{diene}} \quad \text{(} \cdots \text{)}SO_2 \quad \rightleftharpoons \quad \text{\includegraphics[width=1cm]{ring}} SO_2 \qquad (4.3)$$

Approach of a sulphur dioxide molecule to a diene from below and in the plane bisecting the diene leads to favourable overlap of HOMO and LVMO (fig. 4.11), so that a concerted suprafacial, suprafacial addition is allowed. Such reactions are called *linear* cheletropic processes.

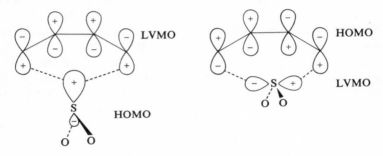

Fig. 4.11. Linear cheletropic addition of sulphur dioxide to butadiene.

Concerted suprafacial, suprafacial addition to an olefin (fig. 4.12) is a disallowed process.

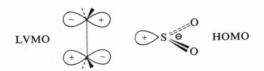

Fig. 4.12

If such a reaction is to be concerted the addition must be suprafacial, antarafacial. This means that in the approach or departure, the atom bearing the lone pair must tip up, so that its orientation with respect to the olefin is different from that in the linear cheletropic process. A transition state geometry which meets the symmetry requirements is shown in fig. 4.13. Here, the axis of the lone pair is perpendicular to the axis of approach or departure. The addition is $_\pi 2_s + _\omega 2_a$; it is equivalent to the $_\pi 2_s + _\pi 2_a$ olefin dimerisation described in § 4.1. Reactions which could involve such *non-linear* addition or extrusion are discussed in § 6.6.

Fig. 4.13

The concerted addition of sulphur dioxide to a triene could formally be either a linear or a non-linear cheletropic reaction. The $6\pi + 2\pi$ addition must be suprafacial, antarafacial. If the sulphur dioxide is the suprafacial component it is a linear cheletropic process and the terminal groups of the triene must rotate in a conrotatory sense, as shown in fig. 4.14(*a*). In the non-linear cheletropic reaction the sulphur dioxide acts as the antarafacial component and the terminal groups must rotate in a disrotatory sense, (*b*). The experimental evidence is that the rotation is conrotatory, so the linear cheletropic pathway is more favourable (§ 6.9).

The retro-additions or extrusions are the microscopic reverse of the forward reactions and follow the same selection rules. These rules are summarised as follows: Linear cheletropic reactions in which the polyene is a suprafacial component (that is, involving disrotatory motion of the termini) are allowed for a total of $(4n + 2)$ electrons. Linear cheletropic reactions in which the polyene is an antarafacial component (involving

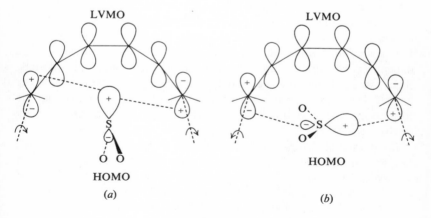

(*a*)

(*b*)

Fig. 4.14

conrotatory movement of the termini) are allowed for a total of $4n$ electrons. The rules are reversed for a non-linear cheletropic change.

Photochemical cycloadditions. Photochemical excitation of an electron to the next highest orbital produces an excited state species of which the HOMO is of opposite symmetry to that of the ground state molecule. It might be expected, therefore, that cycloadditions which are thermally disallowed should be photochemically allowed (one excited reactant molecule adding to an unexcited reactant molecule). In practice, most photochemical cycloadditions appear to be stepwise processes. Many photochemical additions are carried out in conditions where the reacting species is likely to be in a triplet state, either because photosensitisers are used or because the excited singlet undergoes intersystem crossing to the triplet before intermolecular collision with other reactants. It is generally assumed that concerted addition of a triplet molecule is not possible because a reversal of electron spin would be required. On the other hand, intramolecular cycloadditions involving singlet excited species could well be concerted.

With the above points in mind it is nevertheless a useful practical guide that a reaction which cannot be achieved thermally, for example, the formation of cyclobutanes from cycloaddition of two olefins can often be brought about photochemically.

Transition metal catalysed cycloadditions. Certain disallowed cycloadditions take place very much more readily in the presence of transition metal complexes (§ 6.8). The origin of the catalytic effect is still a matter for debate (see chapter 3). Evidence seems to be hardening in favour of stepwise mechanisms for these catalysed reactions.

Retro-cycloadditions. Since retro-cycloadditions are the reverse of the forward reactions the same selection rules apply. It is artificial to separate the forward and reverse reactions because the direction of the reactions depends merely on the relative free energies of reactants and products. Since cycloadditions involve an unfavourable entropy change (from the greater degree of ordering) but a favourable enthalpy change (from the formation of new σ bonds), the reactions can often be reversed by heating.

Retro-cycloadditions can be classified in terms of the σ bonds which are broken. The retro-Diels–Alder reaction shown in fig. 4.15 (*a*) is an allowed concerted $_{\pi}2_s + {_{\sigma}}2_s + {_{\sigma}}2_s$ process; the disallowed mode of cyclobutane cleavage shown in (*b*) is $_{\sigma}2_s + {_{\sigma}}2_s$. The latter reaction would have to be $_{\sigma}2_s + {_{\sigma}}2_a$, as shown in (*c*), in order to be concerted.

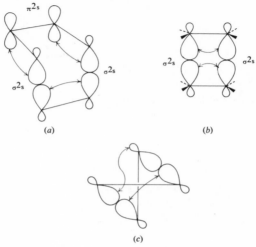

Fig. 4.15

Because of the fundamental importance of the total number of electrons in determining whether a cycloaddition or its reverse can be concerted or not, cycloadditions are discussed in detail in the next two chapters under headings based on the number of electrons involved.

REFERENCES

1. L. A. Paquette, *Principles of Modern Heterocyclic Chemistry*, Benjamin, 1968, p. 337.
2. R. Huisgen, *Angew. Chem. Int. Edn*, 7, 321 (1968). See also R. Huisgen, R. Grashey, and J. Sauer in *The Chemistry of Alkenes*, ed. S. Patai, Interscience, 1964, p. 739.
3. R. B. Woodward and R. Hoffmann, *Angew. Chem. Int. Edn*, 8, 781 (1969).

5 Cycloadditions and eliminations involving six electrons

Thermal cycloadditions and eliminations involving six electrons are by far the most common. Significantly they are in general concerted. This is undoubtedly associated with the allowed nature of concerted $_\pi 4_s + {}_\pi 2_s$ cycloaddition and the fact that such a process is also geometrically favourable.

5.1. The Diels–Alder reaction. The reaction of conjugated dienes with monoenes (fig. 5.1) to give six-membered ring compounds is the best known of cycloadditions and bears the names of the two workers who first recognised the scope of the reaction and began investigation into its mechanism (about 1928).

diene   dienophile

Fig. 5.1. The Diels–Alder reaction.

Scope.[1] Some idea of the scope is given by the examples shown in equations 5.1–5.7. The dienophile can be virtually any π bond (see table 5.1). The diene system is generally all carbon; it may contain hetero-atoms but the structural variation is more limited than for the dienophile (see table 5.2). One or both of the diene double bonds may be part of an aromatic system. The reactions can be carried out in the gas phase or in solution, the rate and temperature required varying with the reactivity of the components. The considerable variation possible in the structure of both diene and dienophile makes the Diels–Alder reaction a very versatile synthetic route to both carbocyclic and heterocyclic compounds. The fact that the retro-Diels–Alder reaction is also general

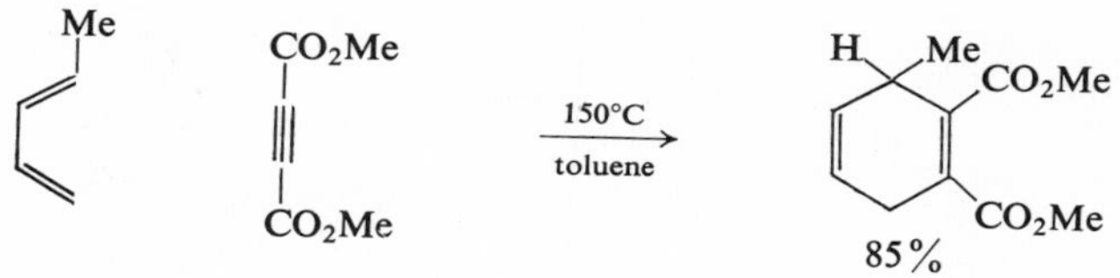

$$\xrightarrow[\text{toluene}]{150°C}$$ (5.1)

85%

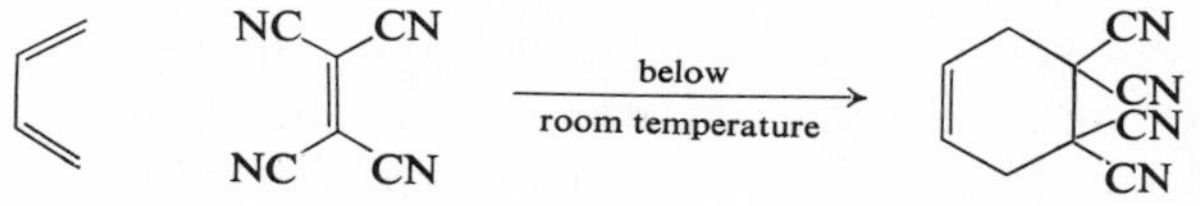

$$\xrightarrow[\text{room temperature}]{\text{below}}$$ (5.2)

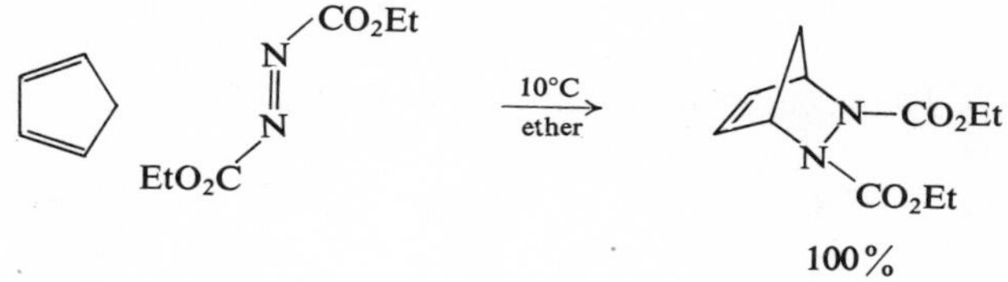

$$\xrightarrow[\text{ether}]{10°C}$$ (5.3)

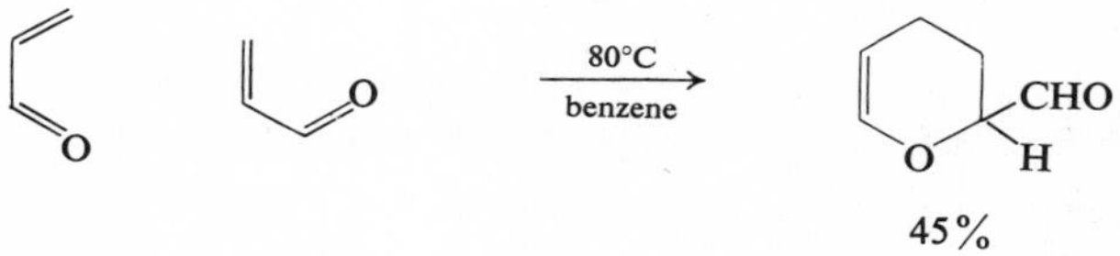

100%

$$\xrightarrow[\text{benzene}]{80°C}$$ (5.4)

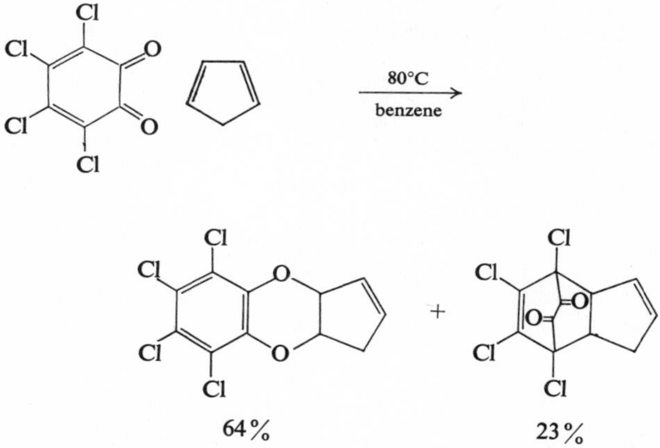

45%

$$\xrightarrow[\text{benzene}]{80°C}$$

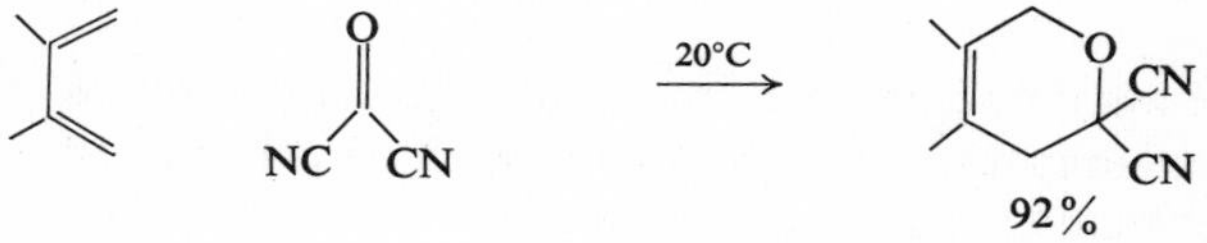

+ (5.5)

64%   23%

$$\xrightarrow{20°C}$$ (5.6)

92%

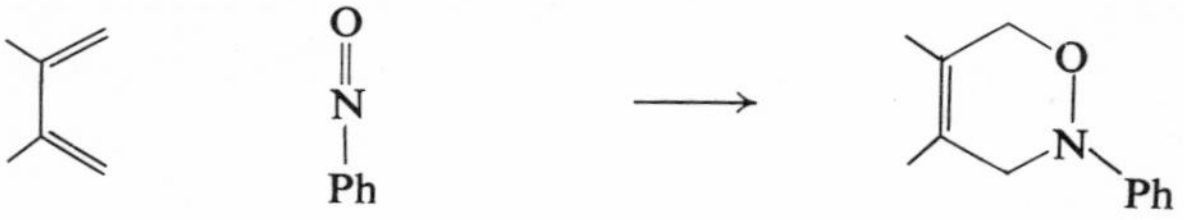

66%

(5.7)

(§ 5.2) considerably extends the synthetic utility of the Diels–Alder reaction since an adduct may undergo a retro-reaction to give components different from those from which it was formed (see § 5.2 for several examples).

TABLE 5.1. π *Bonds acting as dienophiles in Diels–Alder reactions*

>C=C<	—N=N—	
—C≡C—	O=O	(singlet O_2)
>C=N—	—N=O	
—C≡N	N=S	(R—N=S=O)
>C=O	S=O	(e.g. SO_3)
>C=S	Se=O	

TABLE 5.2. *Diene systems acting as enophiles in Diels–Alder reactions*

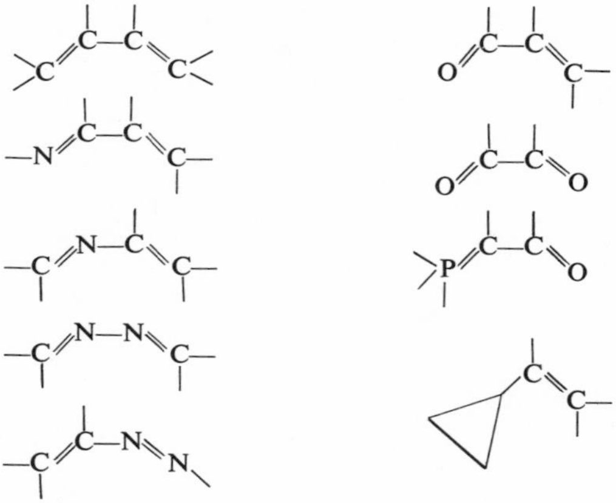

The diene can only react in a *cisoid* configuration; molecules in which the diene system is fused *transoid* do not react, and the rate of reaction with open chain dienes depends on the equilibrium proportion of *cisoid*

conformer present. Thus, substituents in the diene can affect the rate not only by their electronic character but by their influence on conformer proportions. *cis* 1-Substituted butadienes (**1**) are less reactive than their *trans* isomers (**2**) since a bulky R group disfavours the *cisoid* conformation (5.8). Large 2-substituents in the diene favour the *cisoid* conformation and the diene is correspondingly more reactive.

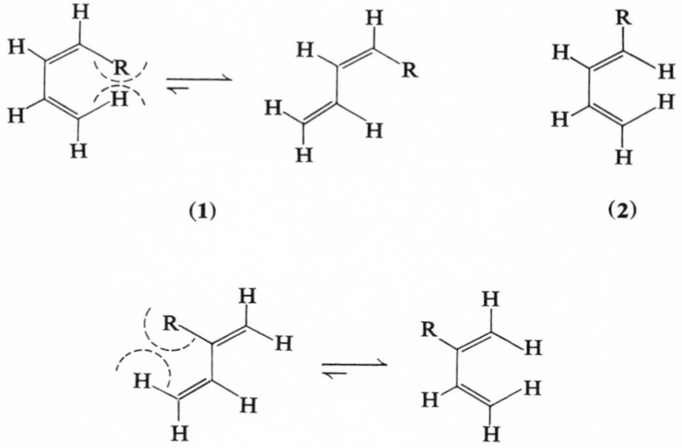

(**1**)   (**2**)

(5.8)

cis Fused cyclic dienes are usually the most reactive, particularly if the ring size is such that the ends of the diene are a convenient distance apart for bonding to the dienophile. Thus cyclopentadiene is somewhat more reactive than cyclohexadiene, both being much more reactive than larger ring dienes, for example, cyclo-octadiene. Cyclic dienoid systems such as furan also take part in the Diels–Alder reaction but as expected the dienoid activity falls in the series furan > pyrrole > thiophen as the diene character is replaced by aromatic character. Pyrroles frequently undergo alternative side reactions and Diels–Alder additions to thiophen have so far only been observed with arynes, which are exceptionally reactive dienophiles.

A similar situation occurs with aromatic hydrocarbons. Benzene acts as a diene only with arynes and highly reactive acetylenes, but naphthalene and particularly anthracene, where there is more 'bond fixation', function more commonly as dienes. Addition to anthracene normally occurs across the 9,10-positions so as to leave two benzenoid aromatic rings (5.9).

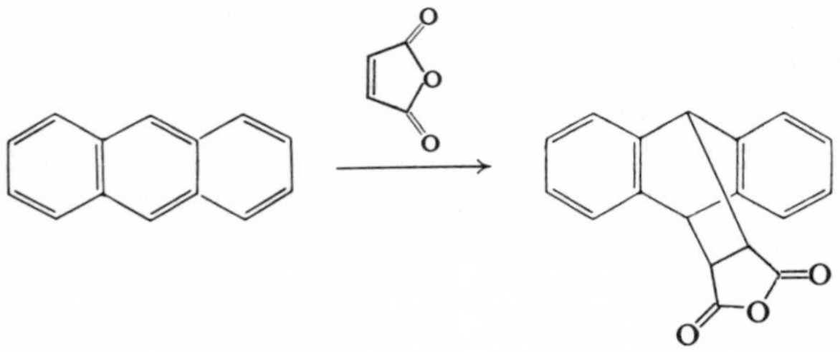

(5.9)

Cyclopentadienones are highly reactive dienes; indeed they are so reactive towards dimerisation (one molecule acting as diene, the other as dienophile) that only highly substituted derivatives, such as the tetra-phenyl derivative (3), are stable as monomers (see p. 127).

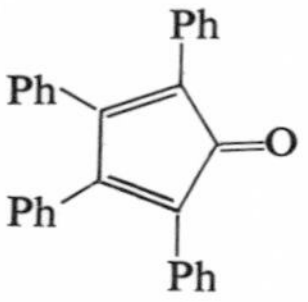

(3)

The common dienes are electron rich and are further activated by electron releasing groups (NMe_2, OMe, Me). It is therefore reasonable that electron deficient dienophiles are the most reactive. Thus olefins or acetylenes bearing electron withdrawing substituents such as acrolein, methyl acrylate, *p*-benzoquinone, styrene, maleate and fumarate esters and acetylenedicarboxylic ester are more reactive than ethylene and acetylene which require strongly forcing conditions. Tetracyanoethylene is a particularly reactive dienophile. The complementary nature of electron rich diene with electron deficient dienophile is very general. With electron deficient dienes however, for example hexachlorocyclo-pentadiene, the electronic requirements of the dienophile are reversed and electron donating substituents favour the reaction. Several examples of such Diels–Alder reactions with so-called inverse electron demand are now known.

Any factor which increases the strain in the dienophile, so raising its energy, increases its reactivity. Thus cyclopropenes are more reactive than corresponding olefins – there is relief of angle strain in the cyclo-propene as the reaction proceeds. Similarly, constraining a triple bond in a ring makes it more reactive. Thus cyclo-octyne is an isolable but reactive dienophile; lower cycloalkynes have not been isolated but their existence has been demonstrated by ready trapping as Diels–Alder adducts.[2] Similarly the reactive intermediate benzyne is very readily intercepted by dienes such as furan, cyclopentadiene or tetraphenylcyclopentadienone.[2] Cyclic azo-compounds with electron withdrawing substituents are highly reactive dienophiles. Unstable species such as (**4**) and (**5**) can be readily trapped and 4-phenyltriazoline-3,5-dione (**6**) is an even more reactive dienophile than tetracyanoethylene.

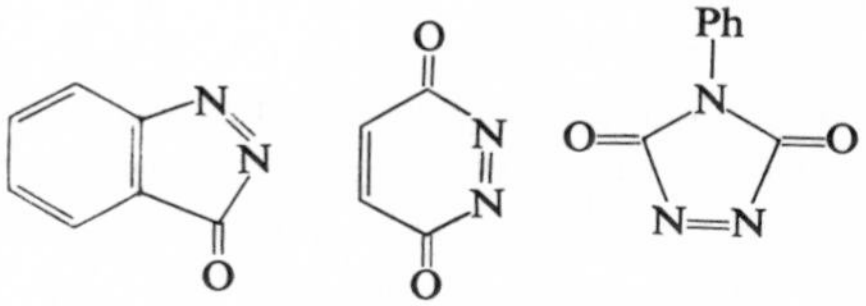

(**4**)   (**5**)   (**6**)

Mechanism.[1a,1c,3,6] Most of the mechanistic work on the Diels–Alder reaction has been carried out with dienes and dienophiles which have all-carbon skeletons. For these cases it was generally recognised, even before the rationalisation of a concerted mechanism by orbital symmetry considerations, that both new bonds were formed at the same time through a transition state (**7**). This is precisely the transition state for $_\pi4_s + _\pi2_s$ addition. Although the two new σ bonds are formed at the same time they are not necessarily formed to the same extent in the transition state. Lack of symmetry in the transition state is more likely to be important where substitution makes the reactants electronically unsymmetrical.

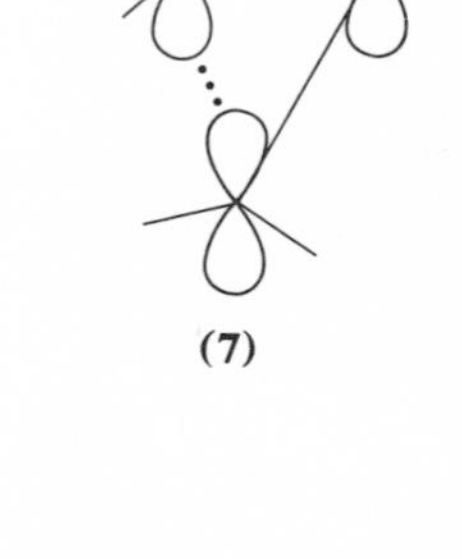

(**7**)

The fact that the reaction also proceeds quite generally when carbon atoms in both diene and dienophile are replaced by hetero-atoms suggests that the same basic mechanism extends to these systems. However, in the absence of detailed mechanistic work, such an assumption may be wrong. The hetero-atoms distort the simple polyene orbitals and in many cases polar intermediates would be highly stabilised. Although some of these reactions may be concerted (with an unsymmetrical transition state) others are almost certainly stepwise. The following mechanistic discussion therefore strictly applies only to the all-carbon Diels–Alder reaction.

The similarity of the rates of the reaction in the gas phase and in solvents of widely differing polarity is hardly consistent with a stepwise reaction through a zwitterionic intermediate. The observed orientation of addition for substituted dienes and dienophiles is also inconsistent with such a mechanism. The effects of substituents on the rate of the reaction, while being very large in absolute terms, are far too small for a process involving a zwitterionic intermediate.

(5.10)

A stepwise reaction through a diradical intermediate can also be ruled out for several reasons. Formation of a diradical intermediate should not be the prerogative of a *cisoid* diene molecule ; *transoid* diene molecules should equally well form such intermediates. However, there is considerable evidence that these intermediates from *transoid* dienes would lead to four-membered ring products (2 + 2 addition). Since most dienes and monoenes only react through the *cisoid* diene molecules (even when the diene is open chain and exists mainly in a *transoid* conformation) to give exclusively six-membered ring Diels–Alder adducts, this is good evidence for a concerted mechanism. (See p. 154 for a full discussion of this point; see also p. 135).

Diels–Alder reactions show high stereospecificity with respect to both diene and dienophile. The specificity is that of $_\pi 4_s + _\pi 2_s$ addition, the relative orientation of substituents on both diene and dienophile being retained in the adduct (5.10).

This is further strong evidence for the concerted reaction since non-stereospecific addition should result from bond rotations competing with ring closure in a diradical or zwitterionic intermediate. The addition of hexachlorocyclopentadiene to α-methylstyrene is a case where a diradical intermediate (**8**) should be particularly favoured since one radical centre would be tertiary and benzylic and the other allylic and stabilised by an α-chlorine. However, even this addition is completely stereospecific.[4]

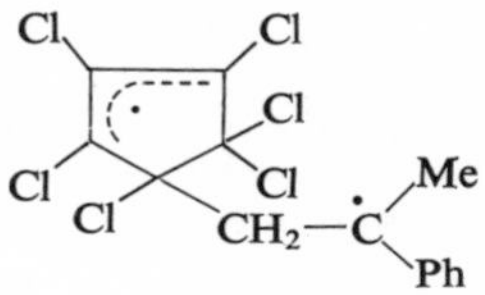

(**8**)

Small inverse secondary H/D isotope effects, $k_D/k_H > 1$, for both diene and dienophile are in agreement with simultaneous but small changes in hybridisation at all four terminal atoms in the transition state. As expected, with highly unsymmetrical dienophiles the isotope effects indicate that although bond formation begins simultaneously it has

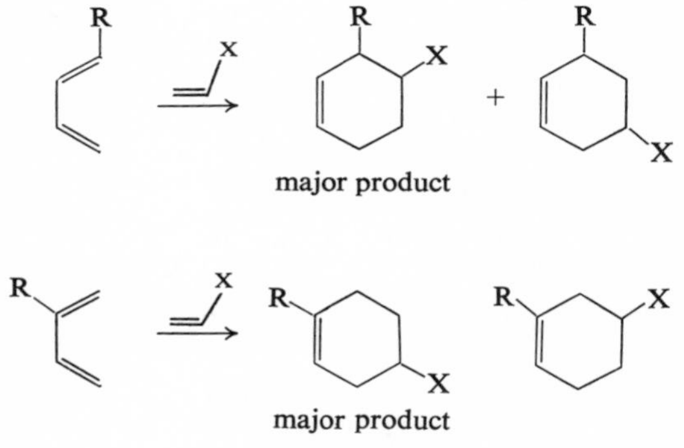

major product

major product

Fig. 5.2

proceeded to different extents in the transition state.[5] The retro-Diels–Alder reaction has frequently been studied to give information of its microscopic reverse – the foward Diels–Alder reaction.[6] Again secondary H/D isotope effects for the retro-reaction of the 2-methylfuran–maleic anhydride adduct indicate synchronous cleavage of both bonds.[7]

Diels–Alder reactions uniformly show large negative entropies of activation in line with the rigid transition state of the concerted mechanism. Thus, the experimental evidence is in line with a concerted $_\pi 4_s + _\pi 2_s$ cycloaddition. However one of the greatest unsolved problems in the Diels–Alder reaction is the question of orientation. No simple explanation is yet available for the orientation of addition of unsymmetrical dienes and dienophiles (see fig. 5.2). The relative yields of structural isomers (fig. 5.2) are not consistent with the reaction proceeding through that transition state in which the build up of charge, due to concerted but unequal bond formation, is best stabilised by the substituents. The orientation is however consistent with reaction through that transition state in which build up of diradical character is best accommodated.

A further important stereochemical consideration in the Diels–Alder reaction is illustrated by the addition of maleic anhydride to cyclopentadiene. Two orientations are possible, that leading to *exo*-addition (9) and that to *endo*-addition (10). In this and other similar Diels–Alder reactions, *endo*-adducts normally predominate, often almost to the exclusion of *exo*-adducts. Initially formed *endo*-adducts may however isomerise to the less sterically hindered, more stable *exo*-adducts by a retro-Diels–Alder reaction followed by recombination (see p. 111).

exo   *endo*

(9)   (10)

Woodward and Hoffmann have put forward an explanation for this widely observed preference for *endo*-addition.[8] Both *endo*- and *exo*-addition is symmetry allowed but for dienophiles with additional π bonds or other available orbitals, secondary orbital interactions operate so as to favour the *endo*-transition state. Thus for cyclopentadiene dimerisation, the *endo*-transition state is stabilised by secondary interactions between appropriate highest occupied and lowest vacant orbitals (fig. 5.3).

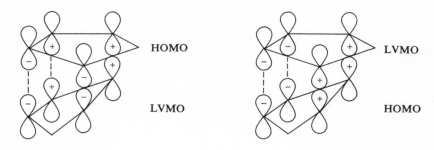

Fig. 5.3

An alternative explanation by Herndon and Hall is that *endo*-addition results from more efficient geometrical overlap for *endo*-addition than for *exo*.[9] This is based on calculations of overlap integrals, for example in cyclopentadiene dimerisation, assuming transition states in which the two molecular planes are parallel for *exo*-addition but inclined at 60° for *endo*-addition. This theory better explains the predominant *endo*-addition of such dienophiles as cyclopropene and cyclopentene (which do not possess the orbitals necessary for secondary interactions). Experiments have been quoted as favouring one or other of these explanations and a clear distinction is not yet possible. Usually, however, the *endo*-preference is greatest in those cases where Woodward and Hoffmann secondary interactions are most readily conceivable. Such secondary interactions can also account for the preferred orientations observed in other cycloadditions[10a] (see p. 195). Other explanations for the *endo*-effect have been put forward. It has been attributed to dipole-induced dipole or charge transfer interactions between the diene and dienophile.[10b]

Although the concerted mechanism is now almost universally accepted, particularly since it fits in so well with orbital symmetry considerations, in the past a considerable amount of work was devoted to investigating the timing of the formation of the new bonds, and inevitably several observations were claimed to support a general stepwise mechanism. Some of the more significant of these will now be discussed.

In certain exceptional cases both 2 + 4 and 2 + 2 adducts are formed in comparable amounts from the same reactants. This has been suggested as support for a common intermediate diradical and hence as support for a radical mechanism for the Diels–Alder reaction, particularly in those cases where the relative proportions of 2 + 2 and 2 + 4 additions did not vary with temperature or solvent.[11] However, it now seems likely that these cases involve competing stepwise radical 2 + 2 addition and concerted 2 + 4 addition, where the two reactants respond similarly to such changes (see p. 157).

Woodward and Katz observed an intramolecular rearrangement of the Diels–Alder adduct (**11**) at 140°C which could be explained by cleavage of bond *a* followed by recombination of the resulting allylic fragments as shown.[12]

(5.11)

(**11**)

Since, at slightly higher temperatures, both bonds *a* and *b* cleave in a retro-Diels–Alder reaction it was assumed that in this retro-Diels–Alder reaction bond *a* cleaved before bond *b*. Thus the microscopic reverse – the forward Diels–Alder reaction was assumed to be a two-stage process. To explain the stereospecificity of the Diels–Alder reactions secondary interactions were assumed to be responsible for preventing rotations about C—C single bonds. This type of argument for a two-stage mechanism has been severely criticised since the rearrangement observed at 140°C is merely a Cope rearrangement (see § 7.4) which need have no connection with the retro-Diels–Alder reaction.[6] Nevertheless the idea of a two-stage mechanism, introduced in 1958 at a time when the one- or two-step nature of the Diels–Alder reaction was still a great problem, provided a valuable compromise between a truly two-step and a one-step mechanism. It implied that the reaction takes place in two stages, in that one bond is largely or completely formed before the other, but it begged the question of whether there is a second energy barrier to be surmounted before closure of the second bond, as there would be in a truly two-step reaction. It explained both the stereochemical characteristics of the reaction and the effect of substituents on the rate and orientation and therefore corresponds to the modern view of a concerted process which can involve an unsymmetrical transition state.

Various determinations of volume of activation for the Diels–Alder reaction have been claimed as support for stepwise mechanisms. However, the theoretical basis of these is uncertain; the most recent determinations indicate a concerted mechanism.

There is thus no evidence to support a *general* stepwise mechanism for the Diels–Alder reaction, although adducts of the Diels–Alder type may be formed by stepwise routes when intermediates are especially stabilised (p. 102).

The concerted $_\pi 4_a + _\pi 2_a$ mode of addition is also formally allowed for the Diels–Alder reaction, but it is geometrically much less favourable than the s,s mode. The a,a mode of addition has tentatively been suggested in a few cases; for example, the thermal rearrangement of octamethylcyclo-octatetraene (**12**) to the semibullvalene derivative (**13**).[13a]

(5.12)

(**12**)   (**13**)

Other mechanistic observations. Charge transfer complexes can be formed between dienes and dienophiles. Colours are often generated on mixing, the colour fading as the adduct is formed. The role of such complexes in Diels–Alder reactions has not yet been elucidated; it may be that the complexes do not lie on the pathway from reactants to products.

Many Diels–Alder reactions can be catalysed by Lewis acids. Such catalysed additions are also highly stereospecific and often show higher orientational selectivity than the uncatalysed reaction, with respect to both *endo/exo*-distribution and the orientation of substituents in the adducts. It seems likely, therefore, that such catalysed additions are basically the same as the uncatalysed; the catalytic action is probably due to complex formation between the Lewis acid and polar substituents on the dienophile components. This increases their electron withdrawing power so making the dienophile more reactive.[3a,13b]

Fukui has suggested that the rates of Diels–Alder reactions depend on the difference in energy between the highest occupied diene orbital and the lowest unoccupied dienophile orbital.[14] Calculations of dienophile reactivity based on these energy differences are found to agree well with the experimental results for the addition of several dienophiles to cyclopentadiene. The ready dimerisation of cyclopentadienones has also been attributed the small energy gap between the HOMO and the LVMO for these systems.[15a] The widely observed accelerating effect of electron releasing groups in the diene and withdrawing groups in the dienophile can be similarly explained.[15b] Electron releasing groups in the diene raise the energy of its HOMO, and withdrawing groups in the dienophile lower the energy of its LVMO, so that the two orbitals interact more strongly.

Stepwise 4 + 2 cycloadditions.[16] Although the majority of 4 + 2 cycloadditions are concerted there are a few cases where zwitterionic intermediates are so stabilised that a stepwise reaction becomes a viable alternative. This applies particularly for heterodienes and heterodieno-philes. Formal Diels–Alder reactions which proceed through zwitterionic intermediates are illustrated by equations 5.13 and 5.14.

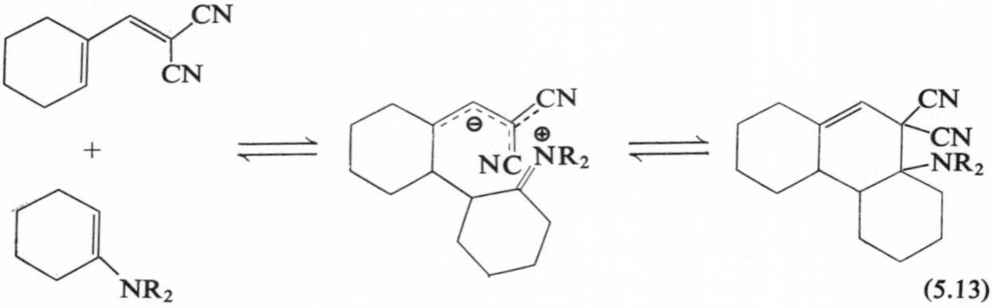

(5.13)

(14)

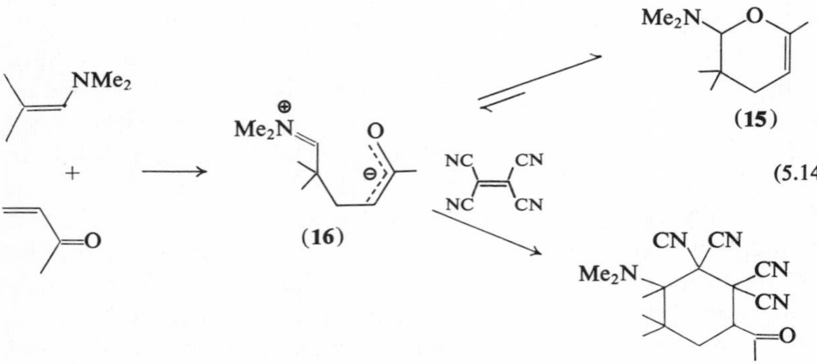

(15)

(5.14)

(16)

In the first example, the zwitterionic intermediate (14) can actually be isolated at −40°C.[16] In the second example, the adduct (15) formed from dimethylaminoisobutene and methyl vinyl ketone is in equilibrium with the zwitterion (16) which can be trapped by tetracyanoethylene.[17]

Thus for $4 + 2$ cycloadditions we can consider a whole spectrum of mechanisms ranging from the fully concerted process, through the asymmetric concerted, to the truly stepwise reaction. In some cases formation of the second bond has become so unimportant in the transition state that it is more realistic to consider the process as stepwise. The vast majority, however, come into the category of symmetrical or unsymmetrical concerted reactions.

There are no clear examples of stepwise $4 + 2$ cycloadditions involving a diradical intermediate (see p. 154 for a full discussion). Diradical intermediates close exclusively or largely to four-membered ring products which are stable under the reaction conditions. Since the zwitterionic reactions, such as those shown above, are readily reversible it may be that the more stable six-ring adducts are isolated because thermodynamic control is operating.

5.2. The retro-Diels–Alder reaction.[6] The retro-Diels–Alder reaction (fig. 5.4) or retro-diene reaction has been known for almost as long as the Diels–Alder reaction itself. Since most Diels–Alder adducts will dissociate if heated sufficiently strongly the scope is comparable with that of the forward reaction.

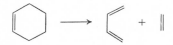

Fig. 5.4. The retro-Diels–Alder reaction.

Retro-Diels–Alder reactions occur more easily when one or both of the fragments are particularly stable. If the diene and dienophile react readily in the forward reaction the retro-reaction is often less favourable. For example, retro-Diels–Alder reactions which involve elimination of nitrogen usually occur very readily, but nitrogen has never been observed to act as a dienophile.

The retro-Diels–Alder reaction has considerable synthetic utility. The prototype reaction, the conversion of cyclohexene into butadiene and ethylene, is used as a laboratory preparation of butadiene.[18] Another important example is the thermal cracking of dicyclopentadiene to give cyclopentadiene (5.15).[19]

Frequently in synthesis the retro-Diels–Alder reaction is used in conjunction with the Diels–Alder reaction. Benzoquinone epoxide cannot be obtained by direct epoxidation of benzoquinone since this

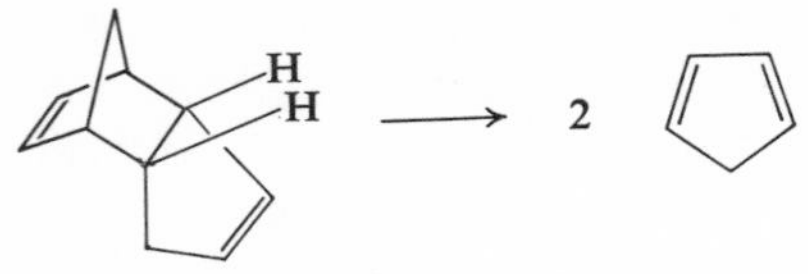

$$(5.15)$$

leads to the anhydride (**17**). However, epoxidation of the Diels–Alder adduct of cyclopentadiene and benzoquinone followed by retro-Diels–Alder reaction of the product provides a route to the mono-epoxide (5.16).[20] A useful preparation of di-imide for reductions involves the Diels–Alder addition of azodicarboxylic ester to anthracene. Hydrolysis of the adduct causes decarboxylation to (**18**) which on heating to ~80°C smoothly decomposes to give di-imide. Reductions can therefore be carried out conveniently in refluxing ethanol (5.17).[21] Vogel's synthesis of benzocyclopropene also makes use of Diels–Alder and retro-Diels–Alder reactions. The cyclodecapentaene (**19**) undergoes Diels–Alder addition via its valence tautomer (**20**) (5.18).[22]

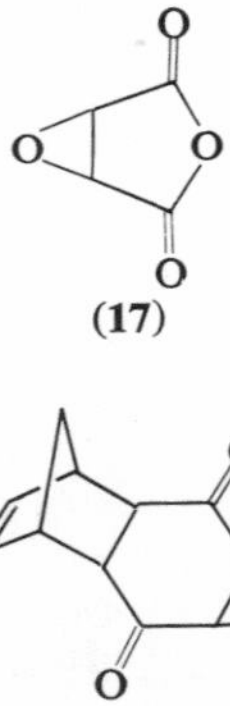

(**17**)

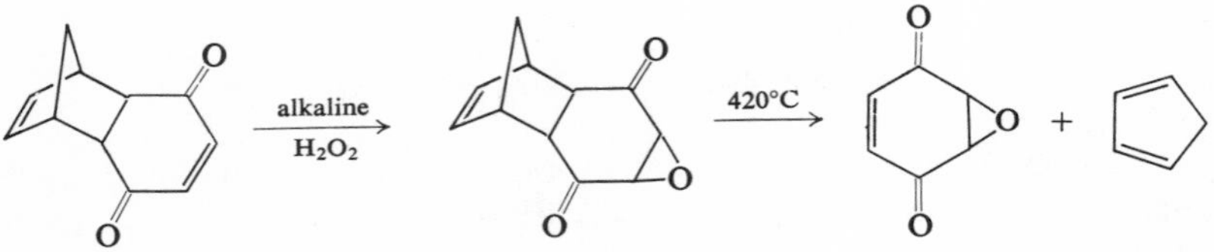

$$(5.16)$$

cis-Dichlorocyclobutene may be formed from cyclo-octatetraene by the sequence illustrated in equation 5.19.[23]

The *trans*-dichlorocyclobutene formed undergoes conrotatory ring opening to 1,4-dichlorobutadiene. However, steric hindrance to conrotatory opening of the *cis*-dichloro-isomer renders it stable.

As can be seen such retro-Diels–Alder reactants are designed so as to ensure that the fragment formed besides the desired product is particularly stable and often aromatic.

The retro-Diels–Alder reaction may be used in conjunction with other cycloadditions as shown by equations 5.20 and 5.21.[6]

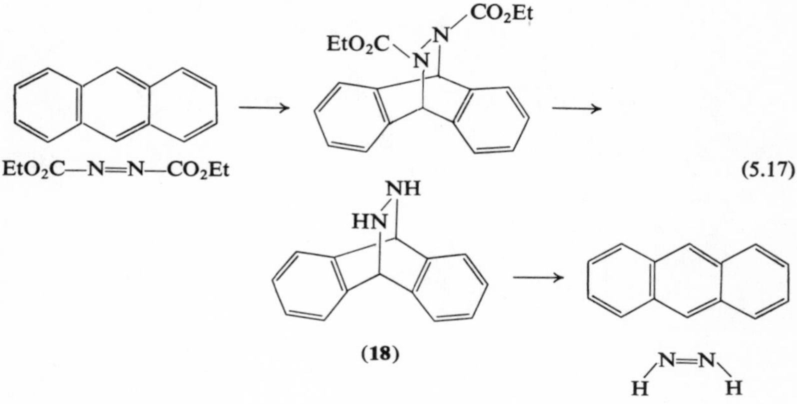

CO2Et
EtO2C
N
N
EtO2C—N=N—CO2Et
NH
HN
(18)
N=N
H
H

(5.17)

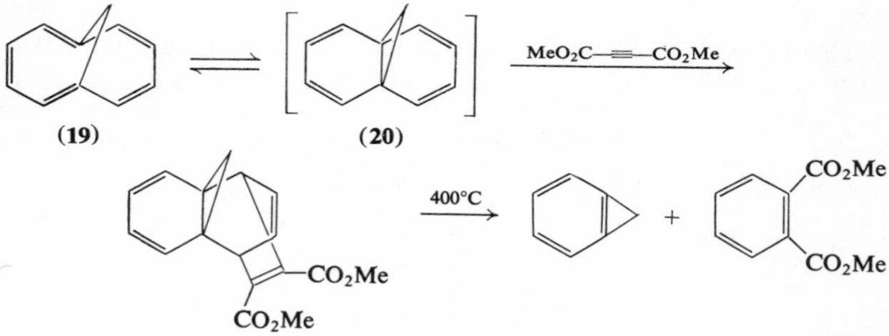

(19)
(20)
MeO2C—≡—CO2Me
400°C
CO2Me
CO2Me
CO2Me
CO2Me

(5.18)

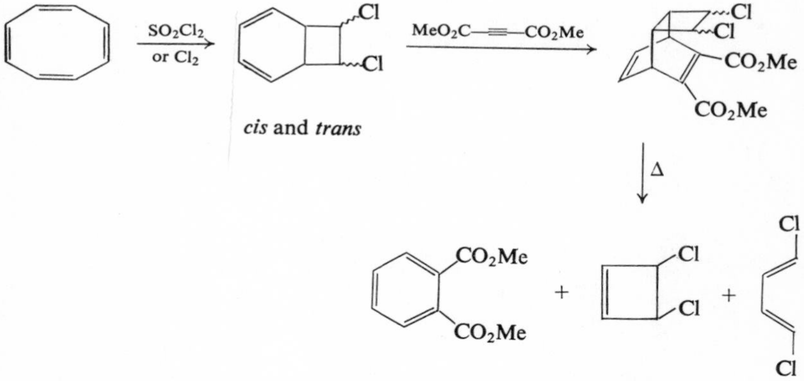

SO2Cl2
or Cl2
Cl
Cl
cis and *trans*
MeO2C—≡—CO2Me
Cl
Cl
CO2Me
CO2Me
Δ
CO2Me
CO2Me
Cl
Cl
Cl
Cl

(5.19)

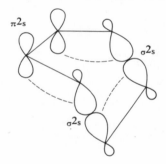

$$\text{Ph-}\overset{\oplus}{\text{C}}\text{=N-}\overset{\ominus}{\text{NPh}}$$

98%

(5.20)

$$\text{H-}\overset{\oplus}{\text{C}}\text{=N-}\overset{\ominus}{\text{O}}$$

135°C

~150°C

90%

(5.21)

The retro-Diels–Alder reaction is the microscopic reverse of the forward reaction; and the bulk of experimental evidence is in support of the expected concerted allowed $_\pi 2_s + _\sigma 2_s + _\sigma 2_s$ process (fig. 5.5).

Fig. 5.5

The reaction is stereospecific as shown in equations 5.22 and 5.23.[24] These particular reactions go smoothly at room temperature. The *cis*

oxidation

(5.22)

(21)

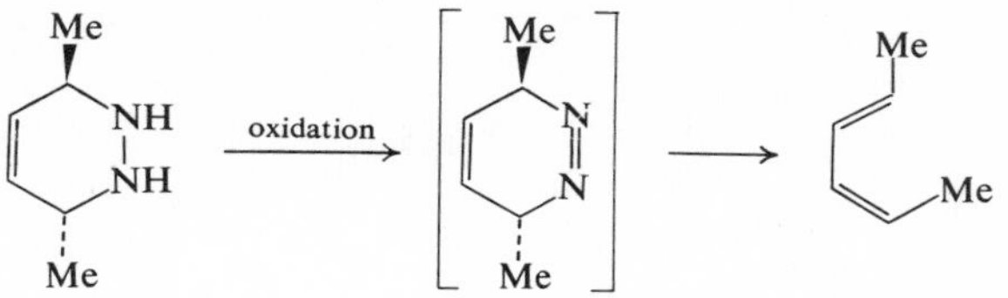

$$(5.23)$$

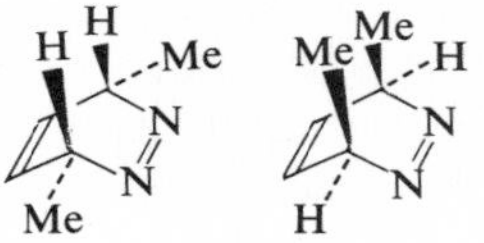

(22)   (23)

isomer (21) only gives the *trans, trans* diene since the conformation necessary for a $_\pi2_s + {_\sigma}2_s + {_\sigma}2_s$ process, leading to *cis, cis* diene, would be too sterically hindered. The reaction takes place through the preferred conformation (22) rather than (23).

The activation entropy for the retro-Diels–Alder reaction is small (+0.5 to −3.5 e.u. in solution, +8.4 to −5.5 in the gas phase; 1 e.u. = 4.18 J mol^{-1} K^{-1}). A stepwise process would involve a large release in ordering and should have a more positive entropy of activation. The small change of ordering is therefore consistent with a concerted process with a cyclic transition state close in character to the adduct.[6] Steric effects of substituents also support this. The forward reaction is greatly retarded by bulky substituents but the retro-reaction is only slightly accelerated as expected if the transition state lies close to the adduct on the reaction co-ordinate.

Isotope effects show that both bonds are cleaved simultaneously although not necessarily at the same rate. Secondary H/D isotope effects for the decomposition of the 2-methylfuran–maleic anhydride adduct (24) indicate equal cleavage of bonds *a* and *b* in the transition state.[7] However primary ^{13}C and ^{18}O isotope effects in the rather exceptional retro-Diels–Alder reaction illustrated in equation 5.24 are consistent with weakening of only bond *a* in the transition state.[25]

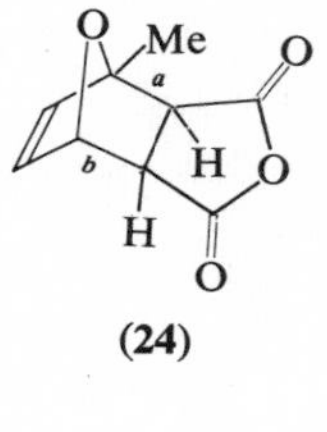

(24)

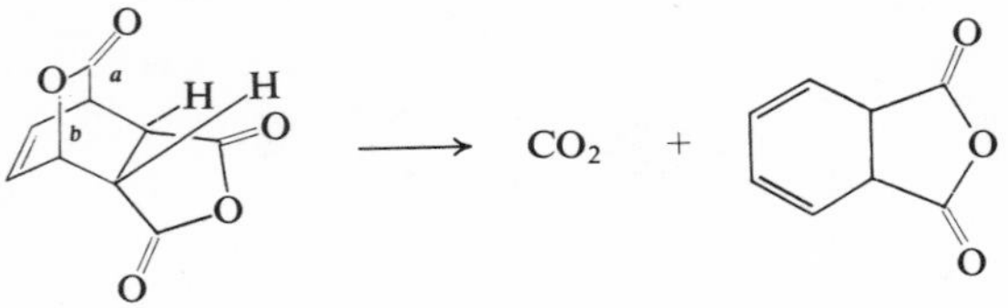

$$\longrightarrow \quad CO_2 \quad + \qquad\qquad\qquad\qquad (5.24)$$

Just as *endo*-adducts are often formed more easily than the *exo*-isomers in the forward reaction so *endo*-adducts undergo retro-Diels–Alder reactions more easily. The transition state for formation and decomposition of *exo*-adducts is higher than that of *endo* (see p. 98 for a discussion of the origin of the effect).

Retro- $_\pi 4_a + {}_\pi 2_a$ addition is also thermally allowed but the unfavourability of this, compared to the normal retro- $_\pi 4_s + {}_\pi 2_s$ mode, is underlined by the behaviour of the two isomers **(25)** and **(26)**.[8] The isomer **(25)**, for which retro- $_\pi 4_s + {}_\pi 2_s$ addition is possible, readily gives butadiene **(5.25)**. However, elimination of butadiene from **(26)** would necessarily either involve a concerted retro- $_\pi 4_a + {}_\pi 2_a$ addition (since the methyl group and hydrogen in the resulting monoene fragment are constrained to be *cis*) or a stepwise fragmentation. Significantly this isomer does not give butadiene even at 400°C.

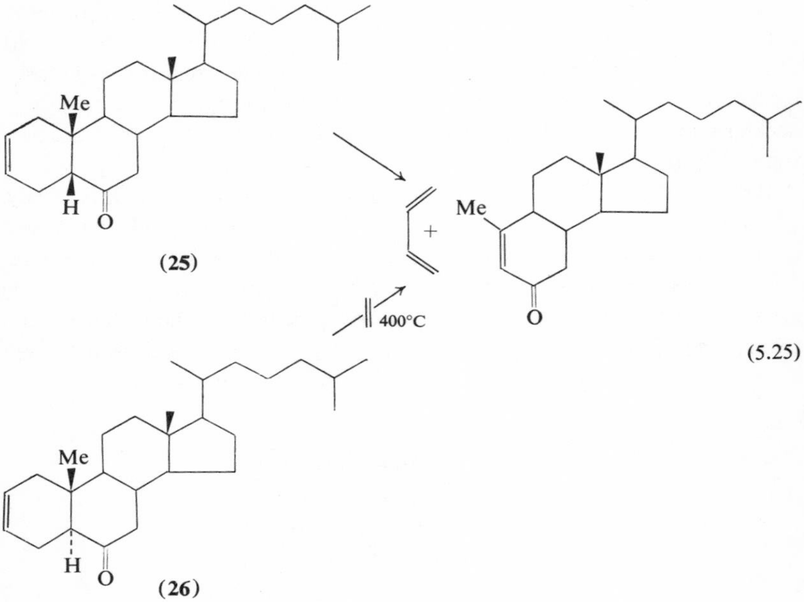

(25)

400°C

(26)

(5.25)

Fragmentation of the cyclic azo-compound (27), which is a retro-Diels–Alder reaction, is much more facile than that of the saturated analogue (28) which does not have the necessary π bond. Compound (27) decomposes as fast as it is formed at −10°C (5.26)[26] whereas (28) only loses nitrogen on heating to 200°C (5.27)[27] In the retro-Diels–Alder reaction as in many other cases, a cyclopropane ring can play the same role as a π bond (see § 7.2). The cyclopropane derivative (29) is intermediate in stability and loses nitrogen smoothly at 25°C (5.28), an estimated 10^{17} times faster than (28).[27,28] Compound (30) also loses nitrogen more than 10^{11} times faster than (31).[29]

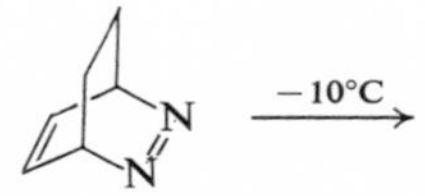

(5.26)

(27)

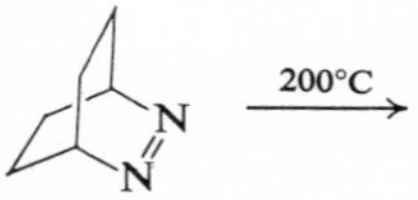

(5.27)

(28)

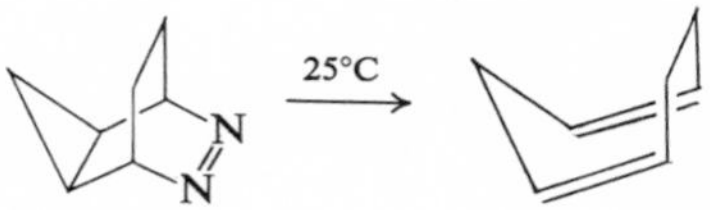

(29)

(5.28)

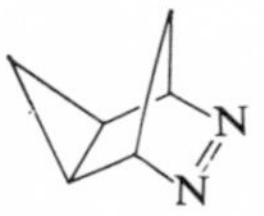

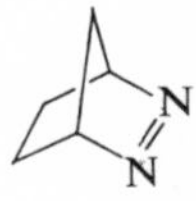

(30)　　　　(31)

Such processes (5.28) are retro-homo-Diels–Alder reactions (for an example of the forward reaction see p. 129). They are also highly stereospecific (5.29–5.31). Nitrogen leaves *anti* to the methylene group of the cyclopropane even though in (32) this involves severe methyl–methyl repulsion.[24] Models show that the necessary orbital overlap is more easily obtained through transition state (33) than (34).

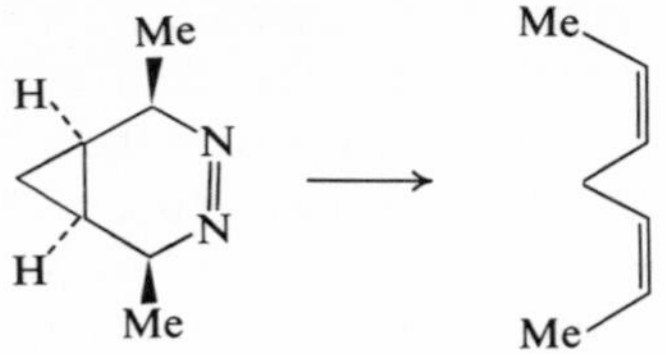

(5.29)

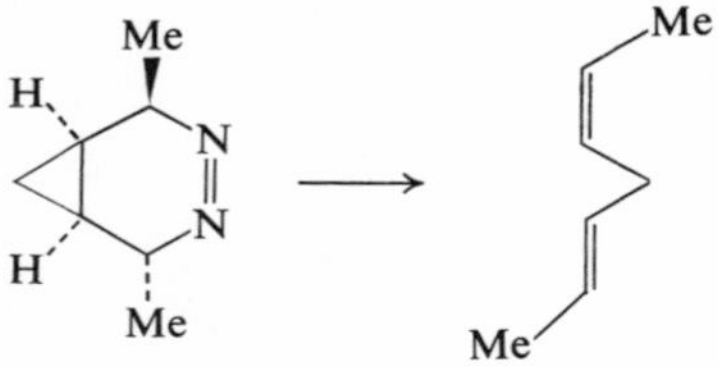

(5.30)

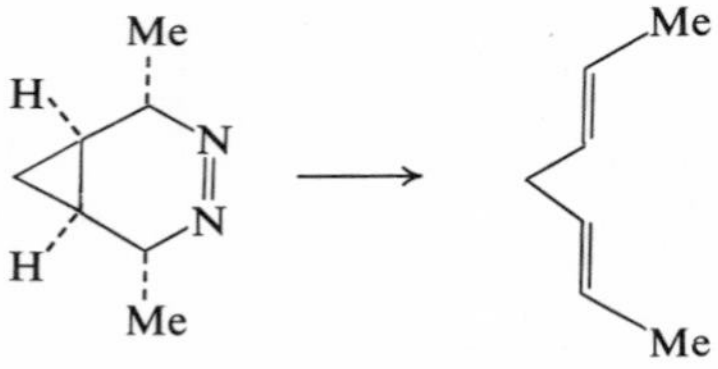

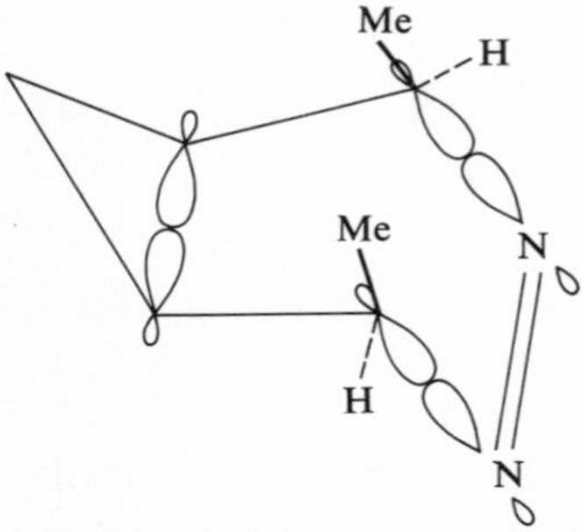

(33)

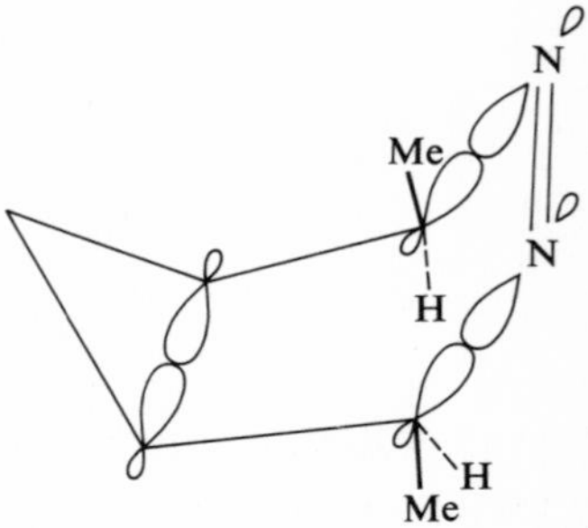

(34)

The mechanism of the *endo-* to *exo*-rearrangement of *endo*-Diels–Alder adducts has received considerable attention. There has in the past been considerable controversy as to whether the rearrangement is intramolecular. However, it now seems clear that it involves a retro-Diels–Alder reaction followed by recombination.[6]

5.3. 1,3-Dipolar cycloadditions (fig. 5.6).[30,1a] In the Diels–Alder reaction, the $_\pi 4_s$ component is a diene in which the four π electrons are distributed over four carbon atoms. An allyl anion has four π electrons distributed over three carbon atoms but the symmetry of the HOMO (fig. 5.7) is the same as that of a diene so that concerted $_\pi 4_s + _\pi 2_s$ addition to a monoene should be allowed.

Fig. 5.6

Fig. 5.7

The correlation diagram for addition of an allyl anion to a monoene is very similar to that for the Diels–Alder reaction, the difference being that a lone pair orbital, rather than a π bond, appears in the product (fig. 5.8).

The simple addition of an allyl anion to a monoene has yet to be observed although a close analogy, the addition of an aza-allyl anion to olefins and acetylenes has been reported (5.32).[31]

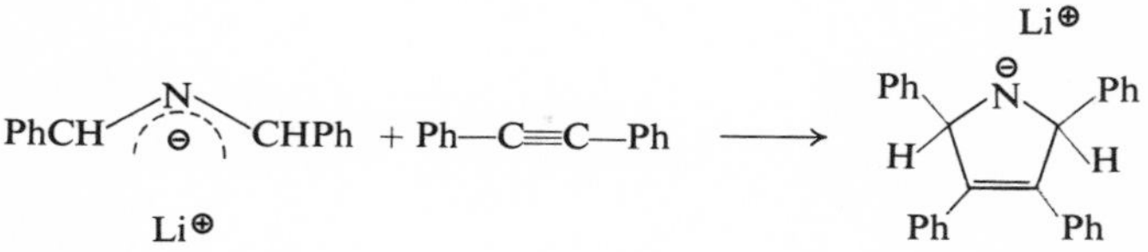

$$\underset{\underset{Li^\oplus}{}}{PhCH\diagup\!\!\overset{N}{\underset{\ominus}{\,}}\!\!\diagdown CHPh} + Ph\!-\!C\!\equiv\!C\!-\!Ph \longrightarrow \underset{\underset{Ph}{}\ \underset{Ph}{}}{\overset{Li^\oplus}{\underset{H}{}\overset{Ph\diagdown\overset{\ominus}{N}\diagup Ph}{}\underset{H}{}}} \qquad (5.32)$$

However, there are numerous examples of the addition of neutral four π electron–three-atom systems to π bonds. These are known as 1,3-dipolar cycloadditions, the 4π system being the 1,3-dipole and the

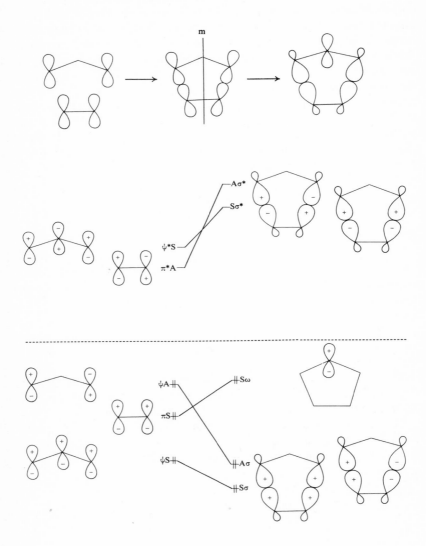

Fig. 5.8. Correlation diagram for addition of allyl anion to monoene (symmetries of orbitals are assigned with respect to the mirror plane m).

2π system the dipolarophile. Although examples (equations 5.33 and 5.34) were known before the turn of the century,[32] such reactions were largely neglected until the late 1950s. Huisgen, in 1958, was the first to recognise fully the general concept and scope of 1,3-dipolar cyclo-addition and since that time, largely due to his efforts, it has become a most valuable method for the synthesis of a great variety of five-ring heterocycles.

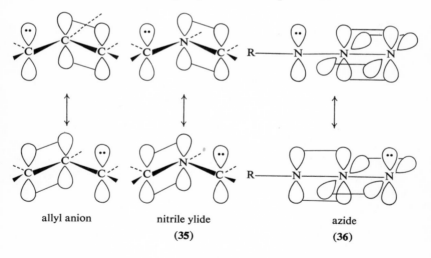

A 1,3-dipole is basically a system of three atoms amongst which are distributed four π electrons as in an allyl anion system (fig. 5.9). The three atoms can be a wide variety of combinations of C, O and N. The dipolarophile can be virtually any double or triple bond.

allyl anion   nitrile ylide   azide

(35)   (36)

Fig. 5.9

The term 1,3-dipole arose because in valence bond theory such compounds can only be described in terms of dipolar resonance contributors, as shown for diazomethane. The extreme 1,3-dipolar forms with their complementary nucleophilic and electrophilic centres readily explain the tendency to undergo addition to π bonds. Indeed it must be possible to write 1,3-dipolar forms for all such species which undergo this type of addition.

$$\overset{\ominus}{:}CH_2\!-\!\overset{\cdot\cdot}{N}\!\!=\!\!\overset{\oplus}{N}\!: \leftrightarrow CH_2\!\!=\!\!\overset{\oplus}{N}\!\!=\!\!\overset{\ominus}{N}\!: \leftrightarrow \overset{\ominus}{:}CH_2\!-\!\overset{\oplus}{N}\!\!\equiv\!\!N\!: \leftrightarrow \overset{\oplus}{CH_2}\!-\!\overset{\cdot\cdot}{N}\!\!=\!\!\overset{\ominus}{N}\!: \leftrightarrow CH_2\!\!=\!\!\overset{\cdot\cdot}{N}\!-\!\overset{\cdot\cdot}{\underset{\cdot\cdot}{N}} \qquad (5.35)$$

$$\textbf{(37)} \qquad\qquad\qquad \textbf{(38)} \qquad\qquad \textbf{(39)}$$

It is important not to misinterpret this picture of 1,3-dipoles. The fact that dipolar forms can be written does not imply a high dipole moment; indeed dipolar forms such as (37) and (39) largely cancel so that such molecules generally have low dipole moments. Diphenyldiazomethane, for example, has a dipole moment of 1.42 D whereas the form corresponding to (38) has a calculated dipole moment of 6 D. It is also impossible to assign a nucleophilic or electrophilic end to the dipole, so it is equally valid to write the cycloadditions as proceeding through either of the 1,3-dipolar resonance contributors (fig. 5.10).

Fig. 5.10

A selection of 1,3-dipoles is presented in table 5.3. They fall into two categories, those without and those with a π bond orthogonal to the 4π allyl system. The former are bent, for example nitrile ylides [fig. 5.9, (35)], and the latter linear, for example azides [fig. 5.9, (36)]. All of the dipoles in table 5.3 are said to have internal octet stabilisation, since the central atom has a lone pair of electrons which is utilised to increase the number of possible resonance contributors (see diazomethane above). The stability of the dipoles varies considerably, for example phenyl azide or ozone can be isolated, but nitrile imines or nitrile ylides can only be generated *in situ*.

The 1,3-dipole may be part of a heterocyclic system. Sydnones (40) and oxazolones (41) are masked azomethine imines and azomethine ylides respectively. They give initial adducts which lose carbon dioxide after rate determining 1,3-dipolar addition (see p. 124 for examples).

TABLE 5.3. *1,3-Dipoles with octet stabilisation*

Name	Structure	Stability and generation
	1,3-Dipoles with an orthogonal double bond	
nitrile ylides	$-\overset{\oplus}{C}=N-\overset{\ominus}{C}\langle \leftrightarrow -C\equiv\overset{\oplus}{N}-\overset{\ominus}{C}\langle$	*in situ* from $\rangle C=N-\underset{Cl}{\overset{H}{\underset{\mid}{C}}}\langle$ and NEt_3
nitrile imines	$-\overset{\oplus}{C}=N-\overset{\ominus}{N}- \leftrightarrow -C\equiv\overset{\oplus}{N}-\overset{\ominus}{N}-$	*in situ* from $\rangle C=N-\underset{Cl}{N}-\underset{H}{\mid}$ and NEt_3
nitrile oxides	$-\overset{\oplus}{C}=N-\overset{\ominus}{O} \leftrightarrow -C\equiv\overset{\oplus}{N}-\overset{\ominus}{O}$	*in situ* from $\rangle C=N-\underset{Cl}{OH}$ and NEt_3
diazo-compounds	$\rangle\overset{\oplus}{C}-N=\overset{\ominus}{N} \leftrightarrow \rangle C=\overset{\oplus}{N}=\overset{\ominus}{N}$	variable stability, isolable in many cases
azides	$-\overset{\oplus}{N}-N=\overset{\ominus}{N} \leftrightarrow -N=\overset{\oplus}{N}=\overset{\ominus}{N}$	stable, isolable
nitrous oxide[a]	$\overset{\oplus}{N}=N-\overset{\ominus}{O} \leftrightarrow N\equiv\overset{\oplus}{N}-\overset{\ominus}{O}$	stable, isolable
	1,3-Dipoles without an orthogonal double bond	
azomethine ylides	$\rangle\overset{\oplus}{C}\diagdown\overset{N}{\diagup}\overset{\ominus}{C}\langle \leftrightarrow \rangle C=\overset{\overset{\mid}{N}\oplus}{\diagdown}\overset{\ominus}{C}\langle$	*in situ* from $\rangle C\diagdown\overset{\overset{\mid}{N}\oplus X^{\ominus}}{\underset{H}{\diagup}}C\langle$ and NEt_3 or electrocyclic opening of aziridines
azomethine imines	$\rangle\overset{\oplus}{C}\diagdown\overset{N}{\diagup}\overset{\ominus}{N}- \leftrightarrow \rangle C\diagdown\overset{\overset{\mid}{N}\oplus}{\diagup}\overset{\ominus}{N}-$	*in situ* from $\rangle C\diagdown\overset{\overset{\mid}{N}\oplus X^{\ominus}}{\underset{H}{\diagup}}N-$ and NEt_3
azoxy-compounds	$-\overset{\oplus}{N}\diagdown\overset{N}{\diagup}\overset{\ominus}{O} \leftrightarrow -N\diagdown\overset{\overset{\mid}{N}\oplus}{\diagup}\overset{\ominus}{O}$	stable, isolable
nitro-compounds[b]	$\overset{\oplus}{O}\diagdown\overset{N}{\diagup}\overset{\ominus}{O} \leftrightarrow O\diagdown\overset{\overset{\mid}{N}\oplus}{\diagup}\overset{\ominus}{O}$	stable, isolable
carbonyl ylides	$\rangle\overset{\oplus}{C}\diagdown\overset{O}{\diagup}\overset{\ominus}{C}\langle \leftrightarrow \rangle C=\overset{O\oplus}{\diagdown}\overset{\ominus}{C}\langle$	*in situ* from electrocyclic opening of epoxides with stabilising groups, e.g.
		$\overset{CN}{\underset{CN}{}}\diagup\overset{O}{\diagdown}\overset{CN}{\underset{CN}{}}$
carbonyl oxides	$\rangle\overset{\oplus}{C}\diagdown\overset{O}{\diagup}\overset{\ominus}{O} \leftrightarrow \rangle C=\overset{O\oplus}{\diagdown}\overset{\ominus}{O}$	reactive intermediate from carbene + O_2
ozone[c]	$\overset{\oplus}{O}\diagdown\overset{O}{\diagup}\overset{\ominus}{O} \leftrightarrow O\diagdown\overset{O\oplus}{\diagup}\overset{\ominus}{O}$	stable

[a] Stable adducts have not been isolated. However, final products can be explained by intermediate formation and decomposition of such adducts.

[b] Very unreactive because of considerable resonance stabilisation. There are no clear-cut examples of nitro compounds acting as 1,3-dipoles.

[c] The widely held view that the first step in ozonolysis of alkenes involves 1,3-dipolar addition of ozone has now been disputed: P. R. Story *et al.*, *J. Amer. Chem. Soc.*, **93**, 3042, 3044 (1971).

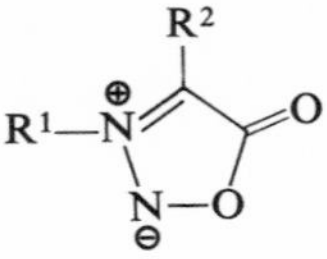

(40)

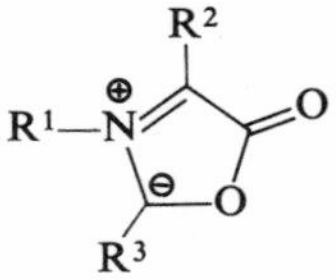

(41)

If the central atom is carbon, octet stabilisation is impossible (see table 5.4). These types of 1,3-dipole are highly reactive species with very short lifetimes, often showing both the reactions of the carbene or nitrene and of the 1,3-dipole. Not all of the possible 1,3-dipolar systems have yet been discovered (table 5.5).

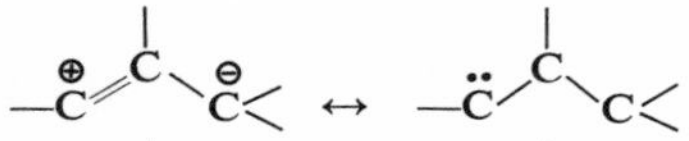

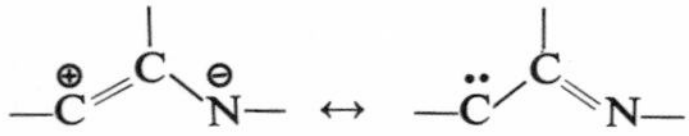

TABLE 5.4. 1,3-*Dipoles without octet stabilisation*[a]

vinyl carbenes	
imino-carbenes	
keto-carbenes	
vinyl nitrenes	
imino-nitrenes	
keto-nitrenes	

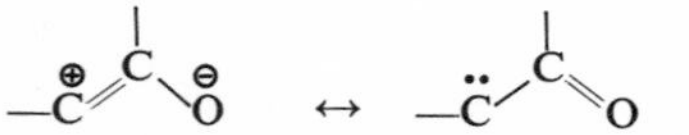

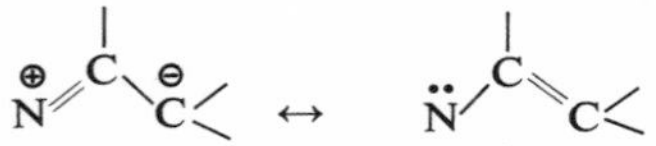

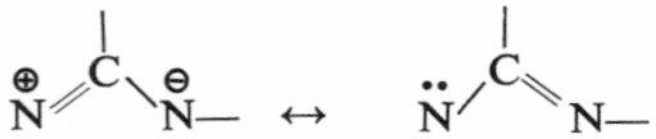

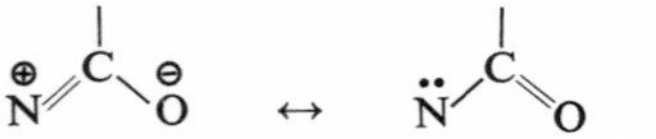

[a] All are highly reactive intermediates.

TABLE 5.5. 1,3-*Dipolar systems not yet discovered*

azimines[b]	
carbonyl imines	
nitroso-imines	
nitroso-oxides	

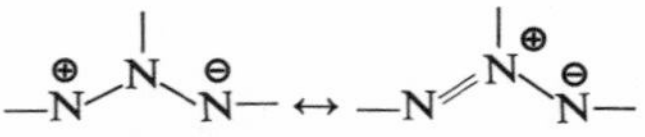

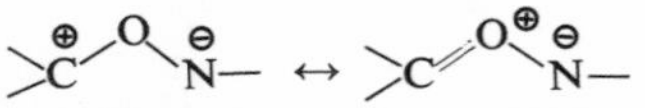

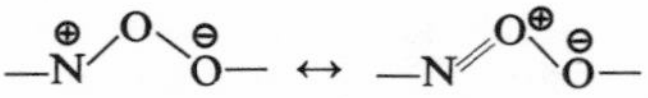

[b] Systems which are formally azimines are known but no cycloadditions have been described.

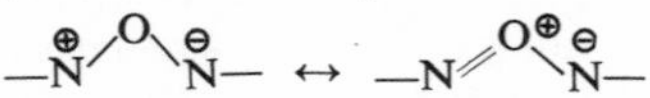

The dipolarophile can be almost any double or triple bond; $\diagdown$C=C$\diagup$,

—C≡C—, $\diagdown$C=N—, —C≡N, —N=N—, $\diagdown$C=O, $\diagdown$C=S, —N=O,

—N=S= (for example in Ph—N=S=O). The π bond may be isolated, conjugated or part of a cumulene system; it can even be the enol form of a ketone or similar compound (5.36).

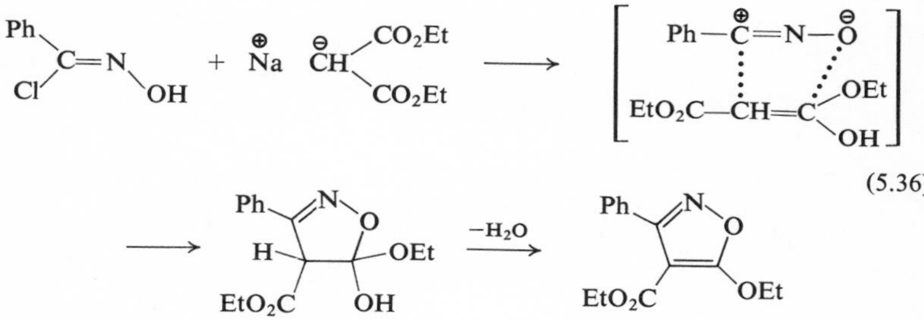

$$(5.36)$$

Olefinic and acetylenic dipolarophiles are more reactive when they bear conjugating groups. Hetero-dipolarophiles are generally less reactive than the corresponding C—C dipolarophiles, possibly because the energy gained from forming the two new σ bonds is less in the former case. This follows because C—C or C–hetero-atom σ bonds are stronger than hetero–hetero-atom bonds (table 5.6).

TABLE 5.6. *Bond energies*

C—O	85 kcal
C—C	83 kcal
C—N	73 kcal
N—N	39 kcal
O—O	35 kcal
N—O	53 kcal

It is the great structural variety of both 1,3-dipole and dipolarophile that makes 1,3-dipolar cycloaddition so valuable and versatile in heterocyclic synthesis. Some additional examples are given in equations 5.37–5.41.

Mechanism.[30,37] Huisgen and others have systematically studied the mechanism of 1,3-dipolar cycloaddition. In most cases the evidence points to a concerted reaction in line with orbital symmetry considerations. The fact that sydnones (**40**) and oxazolones (**41**) behave like other

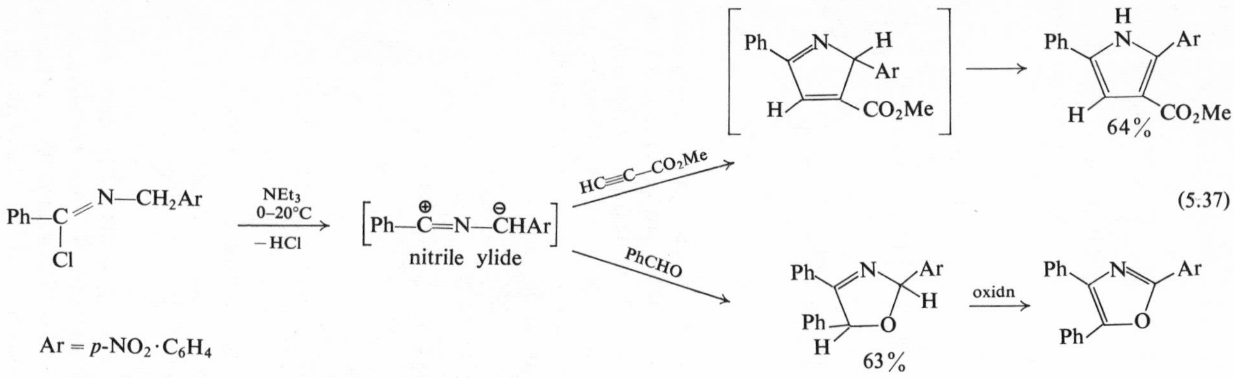

$Ph-C(=N-CH_2Ar)-Cl$

$Ar = p\text{-}NO_2 \cdot C_6H_4$

$\xrightarrow[\;-HCl\;]{\;NEt_3\;0\text{-}20^\circ C\;}$

$\left[Ph-\overset{\oplus}{C}=N-\overset{\ominus}{CHAr} \right]$ nitrile ylide

$HC\equiv C-CO_2Me$

$PhCHO$

64%

63%

$\xrightarrow{\;oxidn\;}$

(5.37)

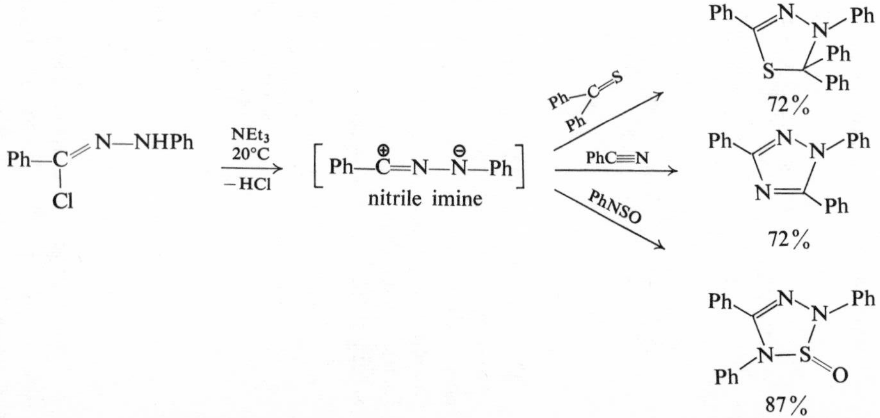

$Ph-C(=N-NHPh)-Cl$

$\xrightarrow[\;-HCl\;]{\;NEt_3\;20^\circ C\;}$

$\left[Ph-\overset{\oplus}{C}=N-\overset{\ominus}{N}-Ph \right]$ nitrile imine

$Ph_2C=S$

$PhC\equiv N$

$PhNSO$

72%

72%

87%

(5.38)

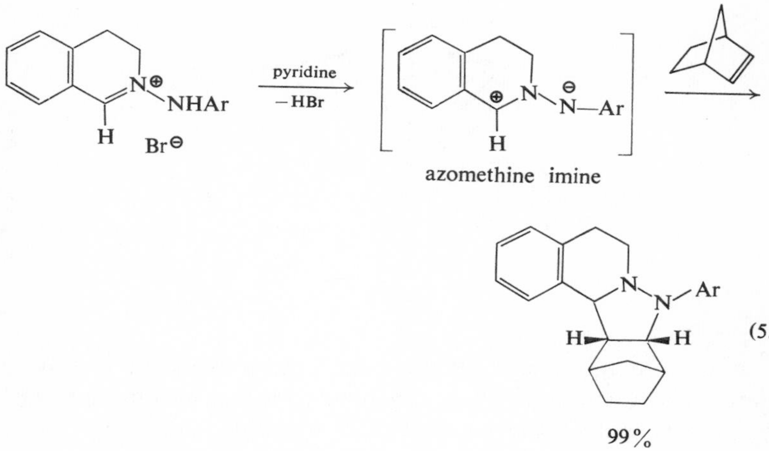

azomethine imine

(5.39)

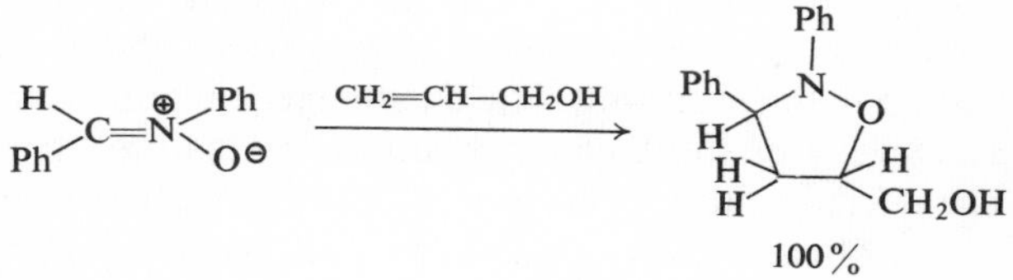

99%

(5.40)

100%

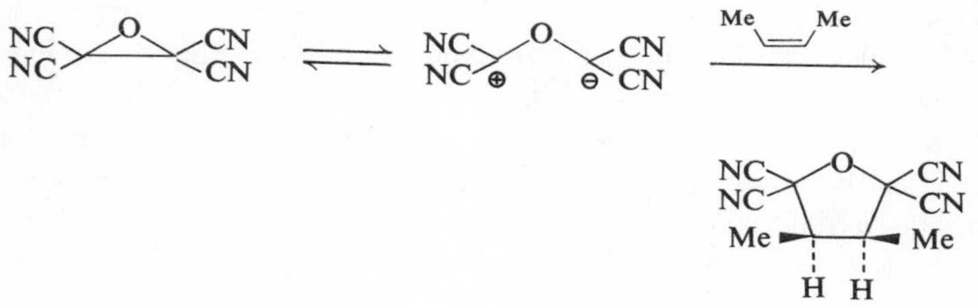

(5.41)

1,3-dipoles is good evidence that the dipole and dipolarophile approach each other in parallel planes ($_\pi 4_s + {_\pi}2_s$) since approach of dipolarophiles to these planar molecules can only be from above or below, as shown in (**42**).

In order to achieve this type of transition state the linear dipoles must first bend; this involves breaking of the orthogonal π bond but leaves the allyl anion π system undisturbed and is not too energy demanding. There is no such problem with the bent dipoles which do not have the orthogonal π bond.

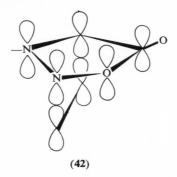

(42)

The rates of 1,3-dipolar additions are largely insensitive to solvent polarity. The reactions have large negative $\Delta S^{\ddagger}$ values and only moderate $\Delta H^{\ddagger}$, and show high stereospecificity in additions to pairs of *cis* and *trans* substituted dipolarophiles. Isotope effects, in the cases studied, indicate a one-step reaction.[33a] Although the formation of the new σ bonds is concerted it is not necessarily symmetrical. Some degree of charge build-up in the transition state due to unequal bond formation is reasonable in view of the highly unsymmetrical nature of most 1,3-dipoles. The effect of substituents in the dipolarophile has been rationalised in terms of their ability to stabilise such partial charges and to increase the polarisability of the dipolarophile π bond. A more attractive explanation is based on the effect of substituents on the energies of the HOMO and LVMO of the reactants and is similar to that discussed for the Diels–Alder reaction (p. 102).[33b] Although most 1,3-dipolar additions are probably concerted, it is quite possible that some are stepwise, since the highly unsymmetrical structure of most dipoles and many dipolarophiles is such that zwitterionic intermediates would be highly stabilised. However, no clear examples of stepwise 1,3-dipolar cycloadditions have yet been reported.

The isolation of open chain hydrazones (**43**) as well as pyrazoles (**44**) in the addition of nitrile imines to aryl acetylenes could be evidence for a stepwise reaction with a common intermediate zwitterion, although the two types of product could be formed independently by separate mechanisms (fig. 5.11).[34] A similar situation exists in the addition of nitrile oxides to arylacetylenes.[35]

As in the case of the Diels–Alder reaction, the orientation of 1,3-dipolar cycloaddition remains a problem. For hetero-dipolarophiles the observed orientation follows that predicted from maximum σ bond formation energy (table 5.6). Although the new σ bonds are only partially formed in the transition state, energy differences for the two possible orientations are sufficiently great to determine the direction of addition. For alkenes and alkynes, the energy differences between opposite orientations are much smaller. In some cases electronic effects appear to be dominant. The direction of addition is such that the substituents can best stabilise the build-up of charge which occurs owing to concerted

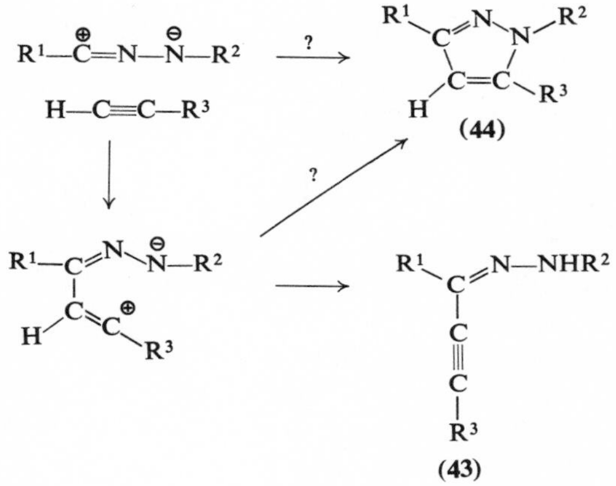

Fig. 5.11

but unequal bond formation in the transition state. In other cases, steric effects which would be expected to be severe in a concerted reaction outweigh such electronic effects. The orientation of these additions is therefore determined by a subtle interplay of steric and electronic factors. The *endo*-effect observed in the Diels–Alder reaction appears to have no parallel in 1,3-dipolar cycloaddition.

An alternative mechanism for 1,3-dipolar additions has been proposed which involves formation of an intermediate singlet diradical.[36] This intermediate rapidly closes stereospecifically if generated in the right conformation for ring closure, but reverts to reactants if not. This theory can be severely criticised on so many grounds that it is not a viable alternative to the concerted mechanism.[37]

Even before Woodward and Hoffmann had formulated their theories, Huisgen had developed a molecular orbital picture of 1,3-dipolar cyclo-addition which is basically the same, and allows one to visualise how the orbitals are transformed.[30b] This is illustrated for additions of the linear diazomethane molecule (fig. 5.12). Initial bending and rehybridisation causes the electrons of the N≡N bond orthogonal to the 4π system to enter a lone pair orbital on the central nitrogen atom. As the interaction between the dipole and dipolarophile proceeds, the system becomes planar and the central nitrogen tips up. As it does so, the central nitrogen lone pair orbital adopts p-character and re-forms the N≡N bond which appears in the product.

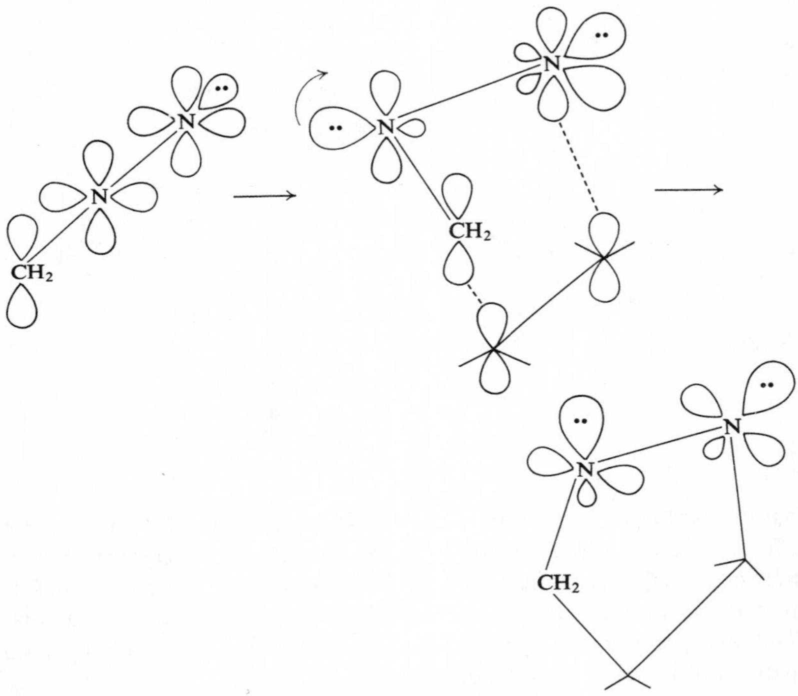

Fig. 5.12

Only two of the four π electrons are required for the formation of the new σ bonds (the dipolarophile provides two) and the other two enter the lone pair orbital on the central nitrogen in the product. The N≡N π bond of the product is therefore formed late in the reaction. This explains why aromatic products are not formed more readily than non-aromatic products in 1,3-dipolar additions; any aromatic stabilisation only comes into play at a very late stage in the reaction and is not therefore appreciable in the transition state.

5.4. Retro-1,3-dipolar cycloadditions.

Retro-1,3-dipolar cycloaddition (fig. 5.13) has not hitherto been recognised as a general reaction class. It is far less well known than the retro-Diels–Alder reaction. Some systems that could in principle undergo this reaction are aromatic and

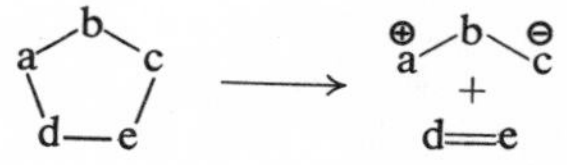

Fig. 5.13

therefore stable; in others, at least one of the fragments (the 1,3-dipole) is not very stable compared with, for example, the fragments from a retro-Diels–Alder reaction. Even so one might anticipate a future growth in the scope of the retro-reaction comparable with that which has taken place for the forward reaction over the last few years.

Several examples of retro-1,3-dipolar cycloaddition do exist. The nitrogen and sulphur ylides (45) and (46), generated by the action of base on the corresponding onium salt, are fragmented as shown in 5.42 and 5.43.[38] Several 1,3-dipolar systems have been generated by thermal

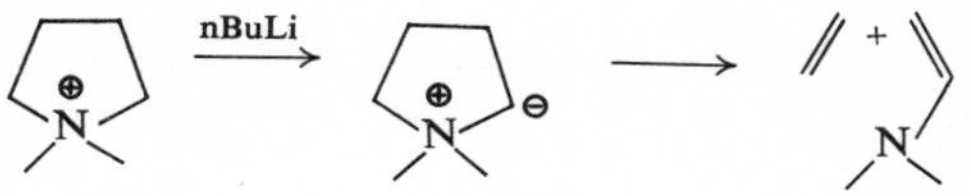

(5.42)

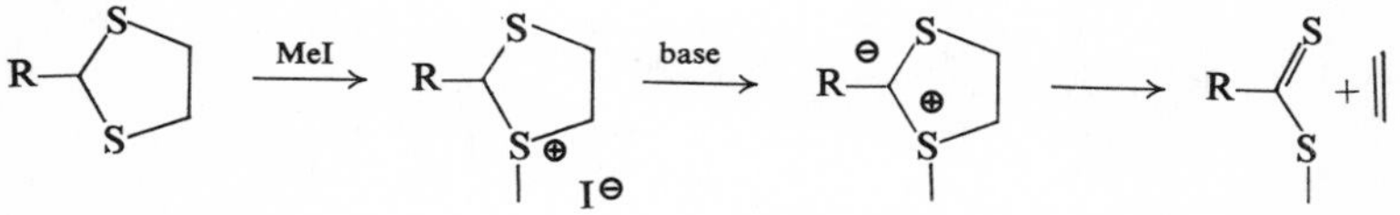

(5.43)

$$Ar\!-\!\overset{\oplus}{C}\!=\!N\!-\!\overset{\ominus}{N}\!-\!Ar + N_2$$

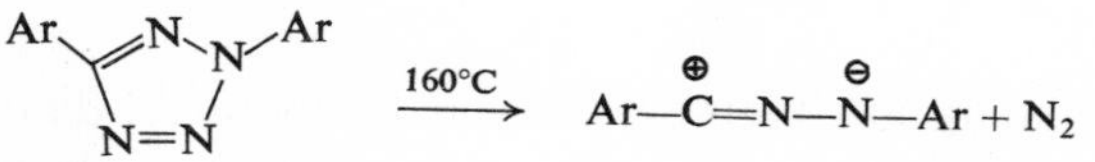

(5.44)

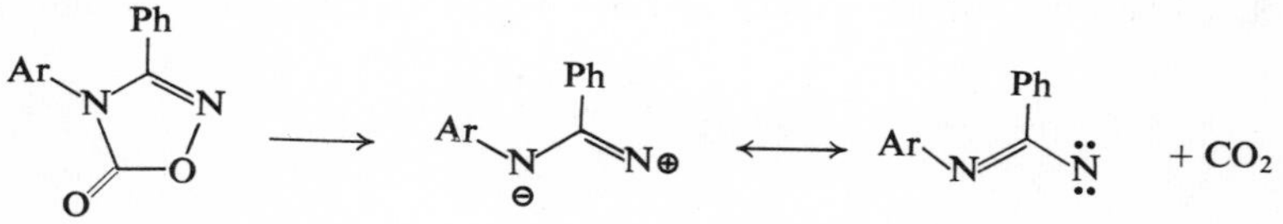

(5.45)

decomposition of heterocyclic systems (equations 5.44 and 5.45). The initial adducts of 1,3-dipoles to sydnones spontaneously lose carbon dioxide in what is formally a retro-1,3-dipolar cycloaddition.[39] The new cyclic 1,3-dipole isomerises or undergoes further reaction to give a stable product (5.46).

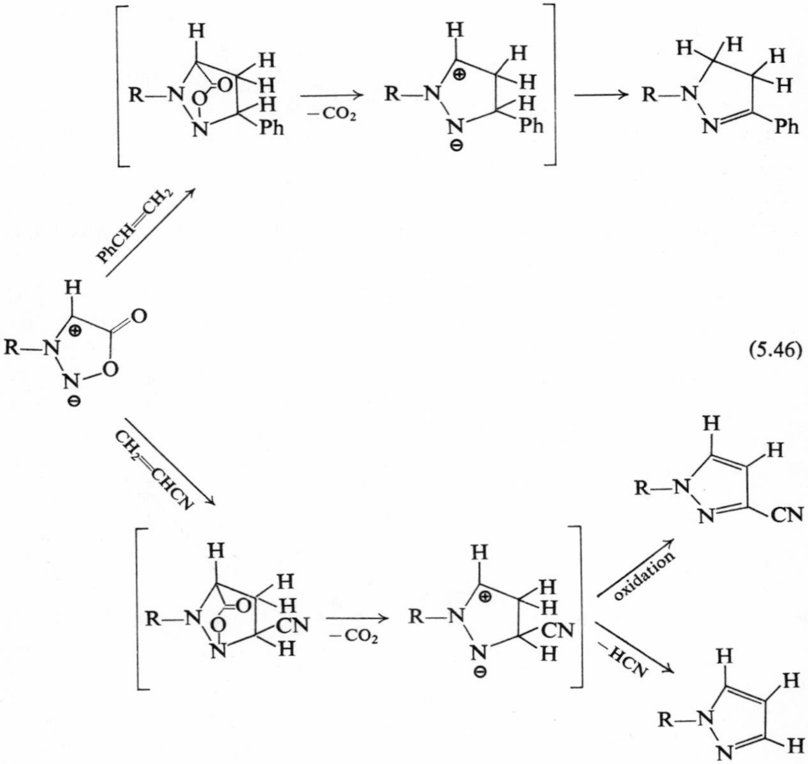

(5.46)

Little mechanistic study has been carried out for retro-1,3-dipolar cycloaddition. However, the concerted process is allowed and would seem likely to operate except in cases where intermediates would be

(47)

highly stabilised. As in the forward reaction, in unsymmetrical cases the two σ bonds may be broken to unequal extents in the transition state. Aryl substituent effects in the thermal decomposition of 2,5-diaryl tetrazoles to nitrogen and diarylnitrile imines (5.44) confirm that the transition state (**47**) is unsymmetrical.[40]

5.5. Cheletropic reactions involving six π electrons.[1d,6,8] Six-electron cycloadditions of the type in fig. 5.14, where both new σ bonds are formed to the same atom of one component, are quite limited in scope. The corresponding retro-cycloadditions (extrusion of X) occur more widely.[41a] This is because X is normally a small inorganic molecule of high thermodynamic stability.

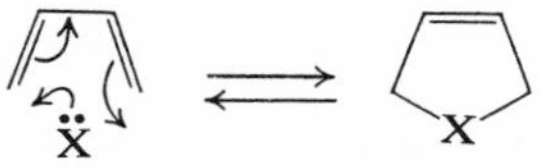

Fig. 5.14

The forward reaction is only well known for sulphur dioxide but even in this case the reverse process takes place readily. This has been put to use in the separation and purification of dienes. The addition of sulphur monoxide has also been reported.[42] Selenium dioxide adds to dienes, but by a Diels–Alder reaction in which the Se=O bond acts as dienophile.[41b] Trivalent phosphorus compounds and dienes give products which can be explained by an initial cheletropic reaction (5.47).

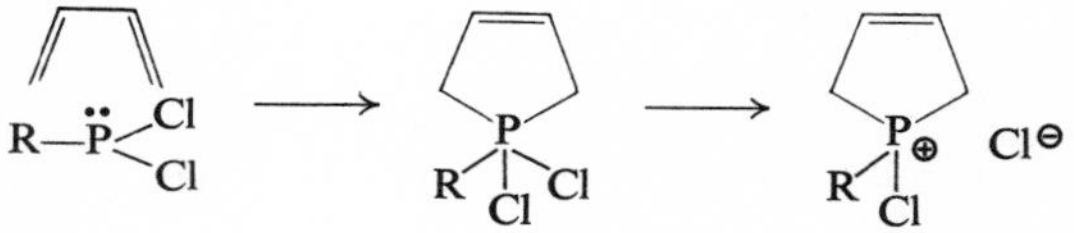

$$(5.47)$$

The elimination (extrusion) reaction is well known for $X = SO_2, N_2$ and CO but is surprisingly inefficient for N_2O (compare the very ready loss of nitrous oxide from *N*-nitrosoaziridines, chapter 6).

The theory of cheletropic reactions has been discussed in chapter 4. These six π electron additions and eliminations are apparently concerted and thus come into the category of linear cheletropic processes ($_\pi 4_s + _\omega 2_s$ cycloaddition and $_\pi 2_s + _\sigma 2_s + _\sigma 2_s$ cycloelimination). They are therefore mechanistically closely related to the Diels–Alder and retro-Diels–Alder reactions, the π orbital of the dienophile being replaced by a lone pair orbital on X.

The thermal addition and extrusion of sulphur dioxide are both stereospecific with disrotatory motions of the terminal methylene groups of the diene (5.48).[43] Photolysis of these sulphones, though not completely stereospecific, is mainly conrotatory as predicted for the photochemical reaction.[44]

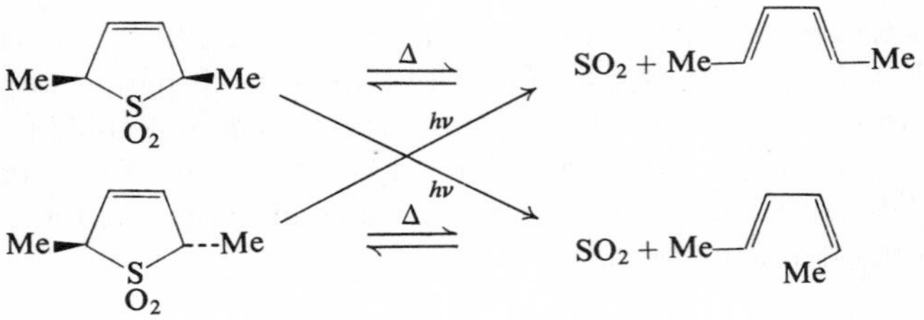

$$(5.48)$$

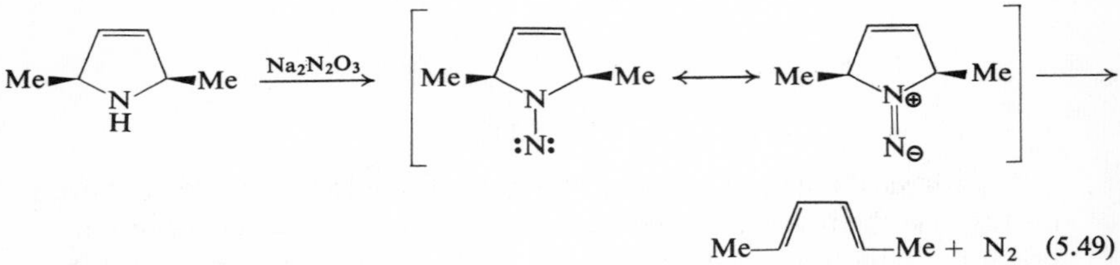

$$Me \diagup\!\!\!\diagdown Me + N_2 \quad (5.49)$$

$$+ N_2 \quad (5.50)$$

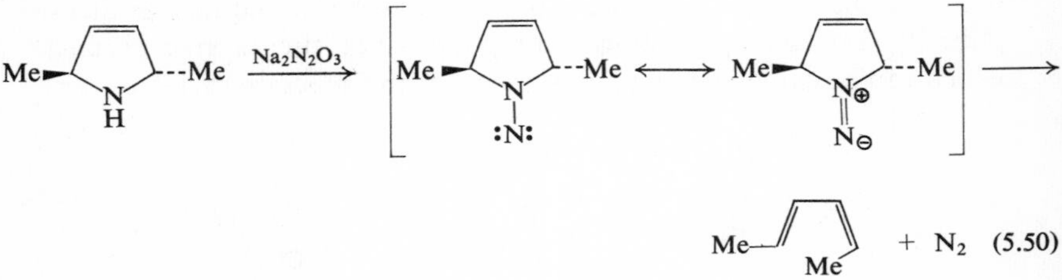

Extrusion of nitrogen from diazenes (equations 5.49 and 5.50) is spontaneous and also stereospecific and disrotatory.[45] Diazenes can be considered as N-nitrenes, and as such can be generated by oxidation of the corresponding N-amino-compounds or by the action of base on the tosyl derivatives. In the example shown the N-nitrenes were generated directly from the NH-compounds by treatment with Angeli's salt $(Na_2N_2O_3)$. Difluoramine (HNF_2) effects the same conversion.

Loss of carbon monoxide from cyclopent-3-enones proceeds much more readily than loss of carbon monoxide from cyclopropanones and

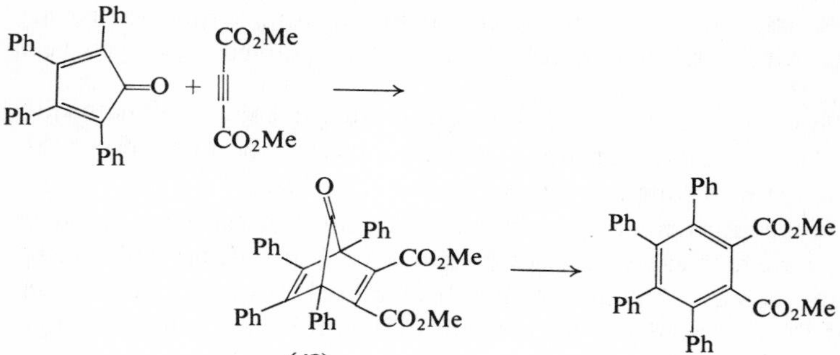

(48)

(5.51)

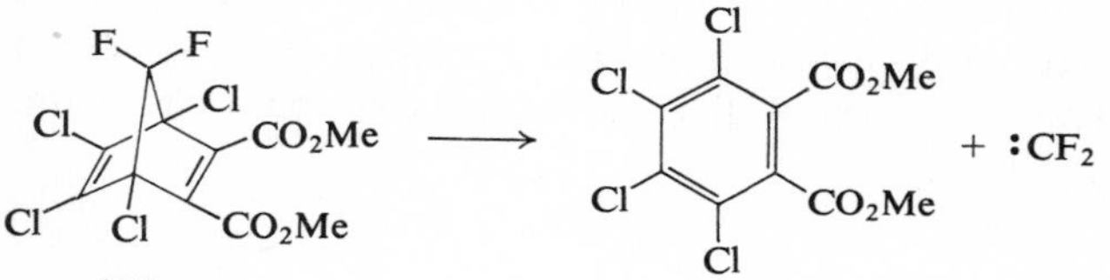

(49)

(5.52)

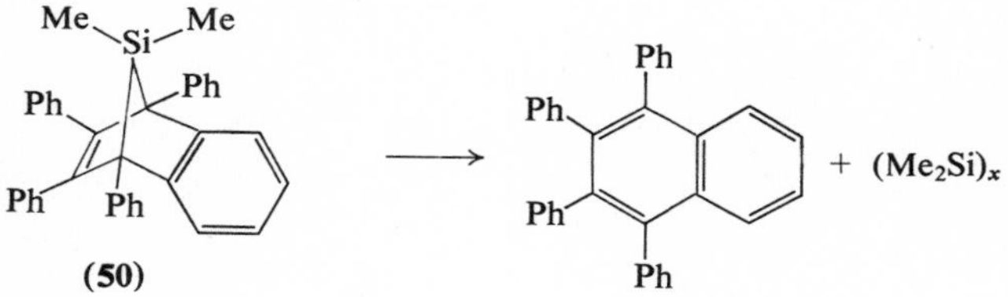

(50)

(5.53)

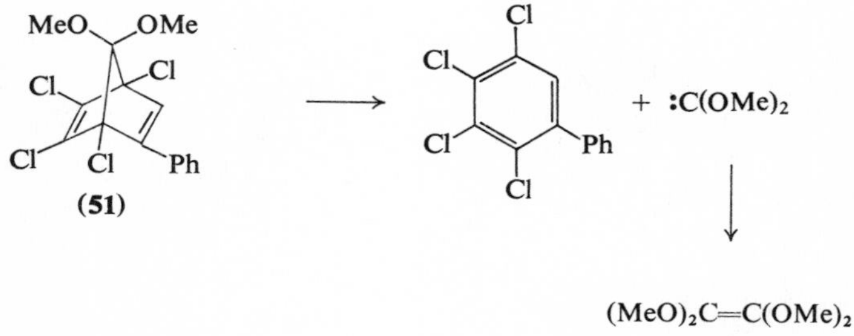

(51)

(5.54)

cyclopentanones. Although no stereochemical studies have been carried out, the generality and ease of the reaction suggests that it is concerted.[46] The loss of carbon monoxide occurs particularly readily when an aromatic product is formed. Thus addition of cyclopentadienones to acetylenes (5.51) and arynes first gives norbornadienone derivatives, for example (**48**), which lose carbon monoxide so rapidly that no primary adducts with the carbon monoxide bridge retained have yet been isolated.[47]

Carbon monoxide is extruded so readily since it is thermodynamically stable. The bridgehead groups in, for example, compounds (**49**) to (**51**) are also lost on strong heating (equations 5.52–5.54).[6]

These reactions could be concerted extrusions of carbenes but since the latter are high energy species, stepwise mechanisms, for which intermediates such as (**52**) would be highly stabilised, must be seriously considered.[48a] Side products, for example (**53**), are formed, which strongly suggest a stepwise breaking of the bridgehead bonds but as yet it is not clear whether such intermediates are involved in the reactions leading to complete loss of the bridgehead group. Decomposition of the ketal (**51**) provided the first synthesis of the interesting electron rich olefin, tetramethoxyethylene.[48b]

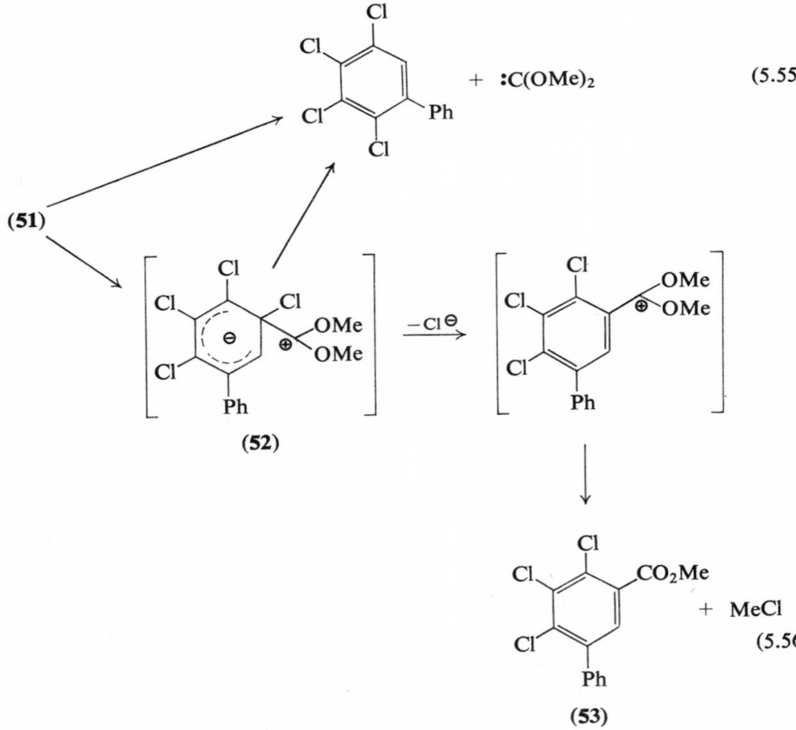

Norbornadienone ketals of type (**54**) are fragmented as shown in (5.57). This may involve concerted or stepwise extrusion of the carbene (**55**) which would be fragmented by a retro-1,3-dipolar cyclo-addition, or the whole process may be concerted.[8]

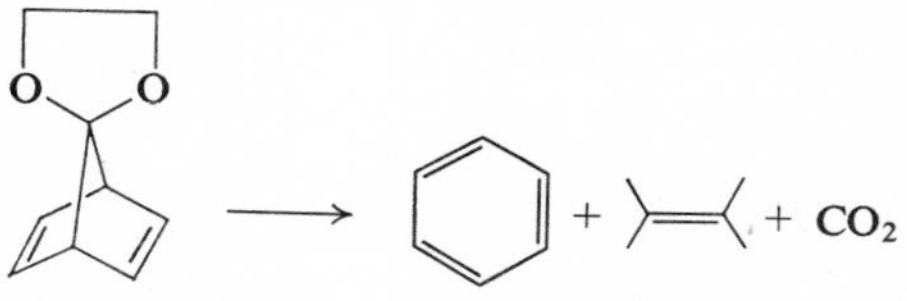

$$(5.57)$$

54

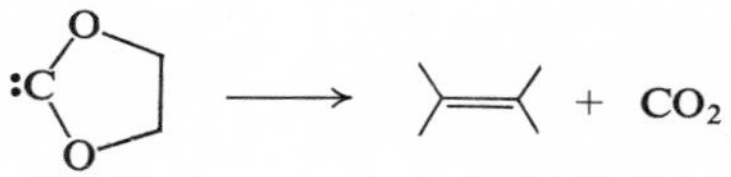

$$(5.58)$$

55

5.6. $2 + 2 + 2$ Cycloadditions and eliminations (fig. 5.15).

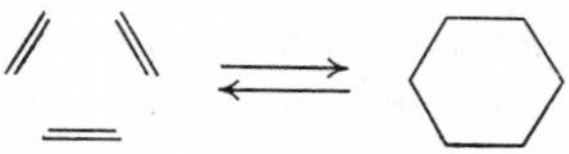

Fig. 5.15

There are many examples of stepwise $2 + 2 + 2$ additions proceeding through 1,4-dipolar intermediates (chapter 6). Concerted $2 + 2 + 2$ cycloadditions are thermally allowed as $_{\pi}2_s + _{\pi}2_s + _{\pi}2_s$ or $_{\pi}2_s + _{\pi}2_a + _{\pi}2_a$ processes. The termolecular collisions necessary for these cycloadditions are very unlikely and the only examples known are those where at least two of the component π bonds are held together in one reactant.

Examples of such $_{\pi}2 + _{\pi}2 + _{\pi}2$ additions are the cycloadditions of tetracyanoethylene and other dienophiles to norbornadiene (5.59) and of tetracyanoethylene to 1,3,5-7-tetramethylenecyclo-octane (5.60).[49]

In both cases the geometry of the molecules favours the process. The first example is often called a homo-Diels–Alder reaction since a cyclopropane ring is formed rather than a π bond.

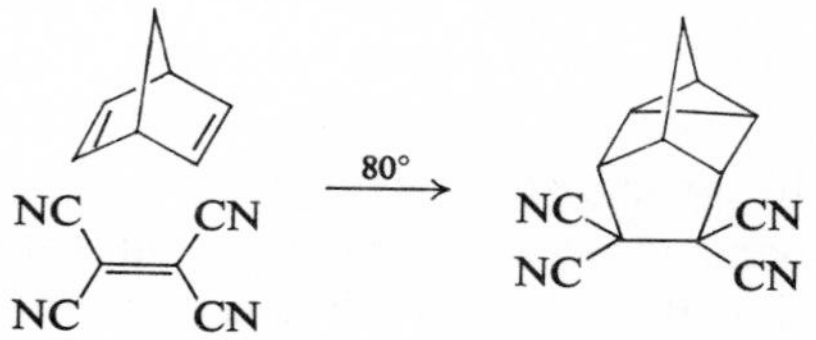

$$(5.59)$$

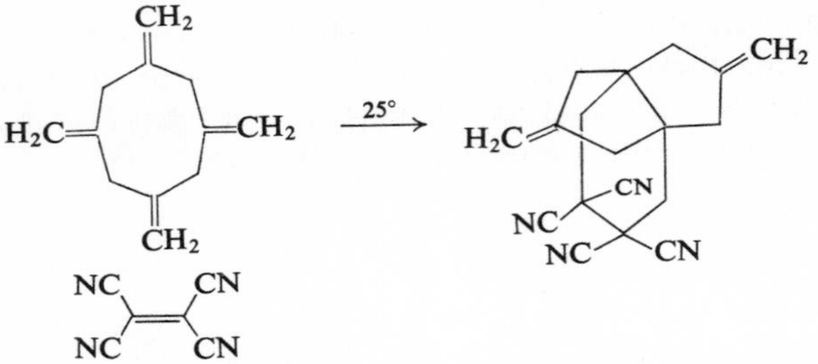

(5.60)

A possible $2 + 2 + 2$ addition involving a cheletropic component is the addition of dichloromethylphosphine to norbornadiene (5.61).[50]

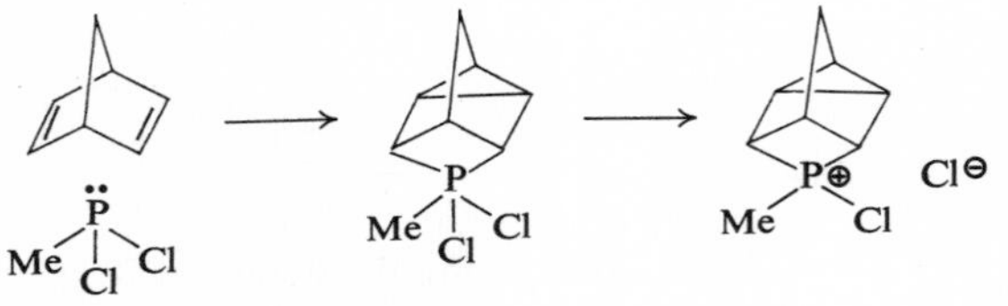

(5.61)

Bis-homo-$2 + 2 + 2$ cycloadditions are also known (5.62 and 5.63).[51,52] The second example is of interest as the first stage in the synthesis of the strained diazetidine discussed on p. 176. This is obtained from the adduct (56) by hydrolysis and decarboxylation followed by oxidation.[52]

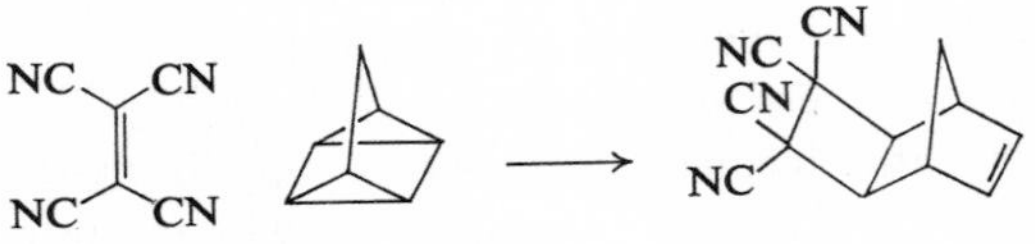

(5.62)

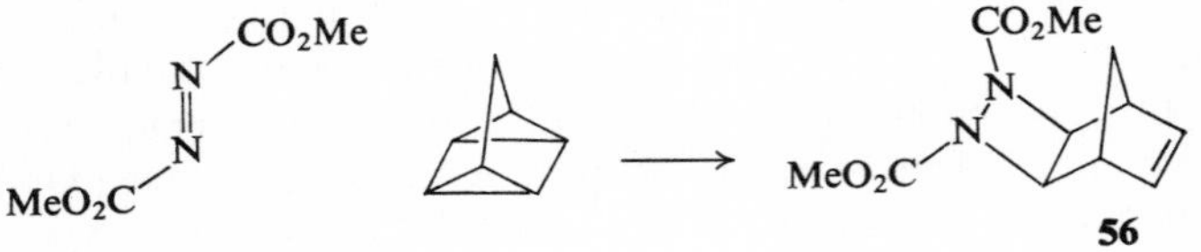

(5.63)

56

Concerted retro-$2 + 2 + 2$ cycloadditions ($_\sigma 2_s + _\sigma 2_s + _\sigma 2_s$ or $_\sigma 2_s + _\sigma 2_a + _\sigma 2_a$ fragmentations) do not suffer from the highly unfavourable entropy requirements of the forward reactions. The concerted retro-homo-Diels–Alder reaction (p. 109) can be considered as a retro-$2 + 2 + 2$ cycloaddition. Perhaps surprisingly, few simple retro-$2 + 2 + 2$ additions leading to three fragments have been reported. One example is the stereo-specific fragmentation of the dihydro-oxadiazinone (57, equation 5.64).[53]

Another example is the thermal fragmentation of the azo-lactone (**58**, equation 5.65) to diethylketen, cholestanone and nitrogen.[54] It is interesting to note that the photo-induced fragmentation of (**58**) takes an alternative course and is probably stepwise.

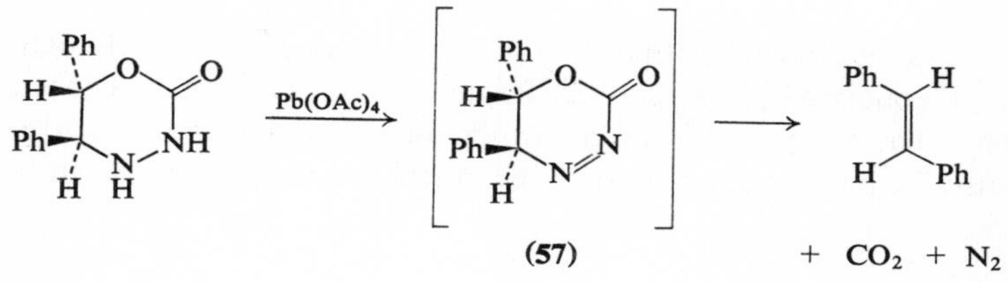

(**57**)

$+ \quad CO_2 + N_2 \quad (5.64)$

Δ
115°C

$+ \quad {}^{Et}_{Et}C{=}C{=}O \; + \; N_2$

(5.65)

$h\nu$

$+ \quad CO_2$

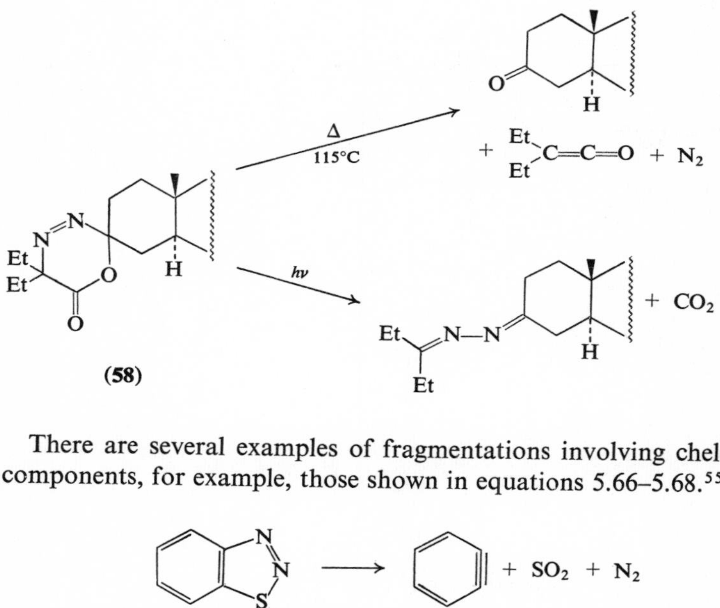

(**58**)

There are several examples of fragmentations involving cheletropic components, for example, those shown in equations 5.66–5.68.[55]

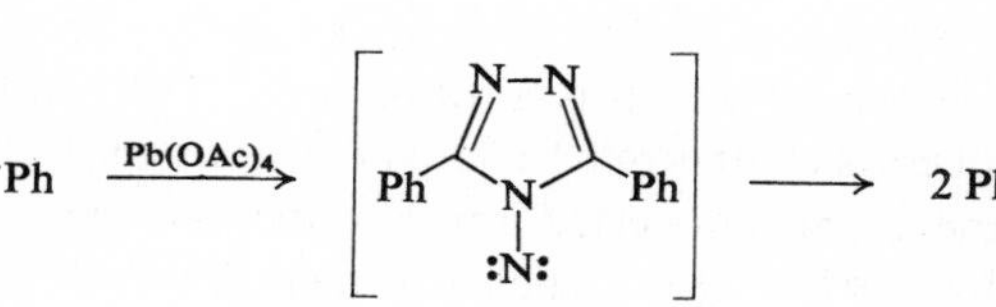

$+ \; SO_2 \; + \; N_2 \qquad (5.66)$

$\xrightarrow{Pb(OAc)_4}$

$\longrightarrow \quad 2\,PhCN \; + \; N_2$

(5.67)

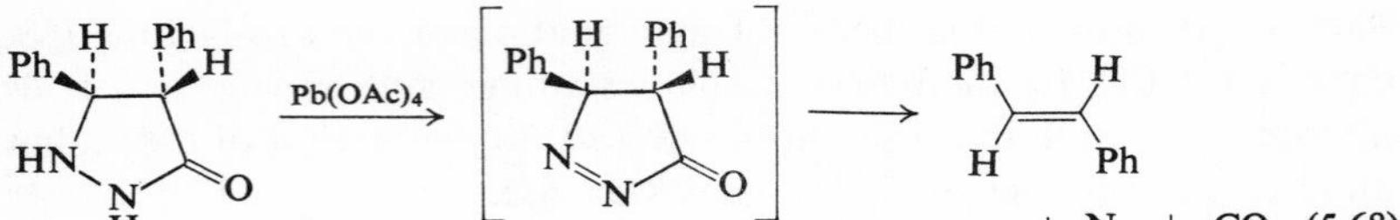

$$+ N_2 + CO \quad (5.68)$$

The rearrangement of prismane to benzene could geometrically take place by a concerted $_\sigma2_s + _\sigma2_s + _\sigma2_s$ fragmentation. One might expect, therefore, that the transformation would be allowed and that a tremendously strained molecule such as prismane would very readily

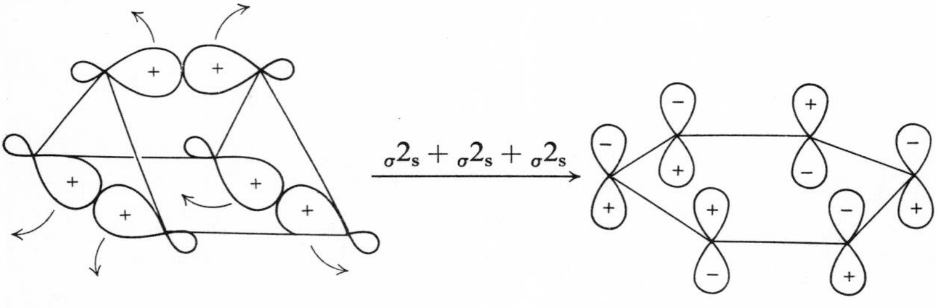

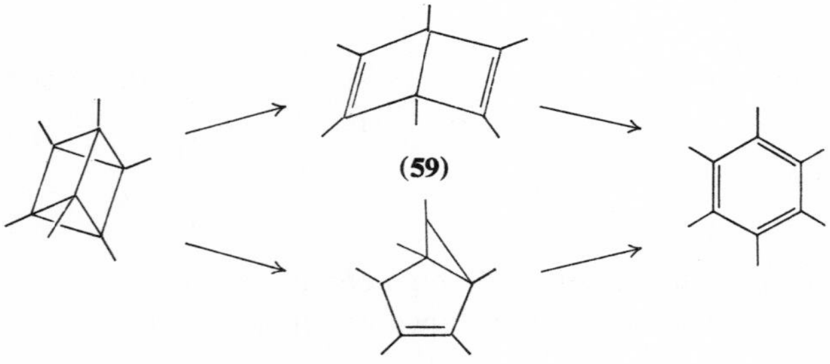

(59)

(60)

Fig. 5.16

isomerise to benzene. However, the π components resulting from the $_\sigma2_s + _\sigma2_s + _\sigma2_s$ cleavage are held together in a ring by σ bonds and actually form a benzene orbital which is antibonding.[8] This explains why, although the process is exothermic by 91.2 kcal mol^{-1} (381 kJmol^{-1}), hexamethylprismane is only converted into hexamethylbenzene above 60°C, and then not directly but via the valence isomers (59) and (60) (fig. 5.16).[56]

5.7. Other six-electron cycloadditions. This section covers some obscure but theoretically interesting reactions.

The $_\pi 4_s + _\pi 2_s$ additions of allyl anions have already been discussed. The symmetry of the HOMO of an allyl cation is such that it should undergo concerted $_\pi 4_s + _\pi 2_s$ addition to a diene. This has been observed in the addition of the methylallyl cation (from $Ag^\oplus$ catalysed ionisation of 2-methylallyl iodide) to cyclopentadiene, cyclohexadiene and furan

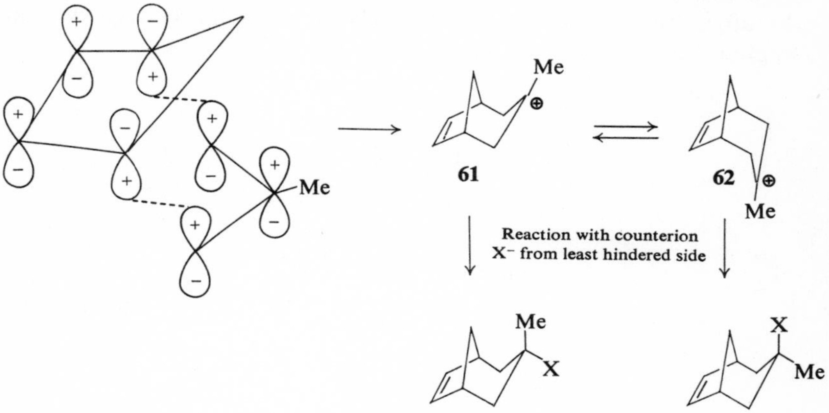

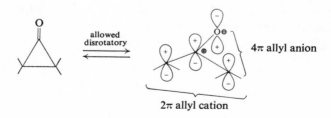

Fig. 5.17

(fig. 5.17). Secondary orbital interactions indicate that *exo*-addition should be most favourable. This appears to be the case but the picture is complicated since the initially formed *exo*-cation (**61**) can 'flip' to the *endo*-form (**62**). Under conditions where the adduct cation has a short

Fig. 5.18

lifetime and reacts rapidly with its counter-ion, only products derived from the *exo*-ion are formed. Other products result from proton loss from the two cations.[57]

Cyclopropanones undergo cycloadditions, apparently through an equilibrium proportion of ring opened form. This ring opened form has the orbital characteristics of both a 2π allyl cation and a 4π allyl anion (fig. 5.18).

$4\pi + 2\pi$ Additions both of the allyl cation system to dienes and of the allyl anion system to the π bonds of trichloroacetaldehyde and sulphur dioxide have been observed (fig. 5.19).[58]

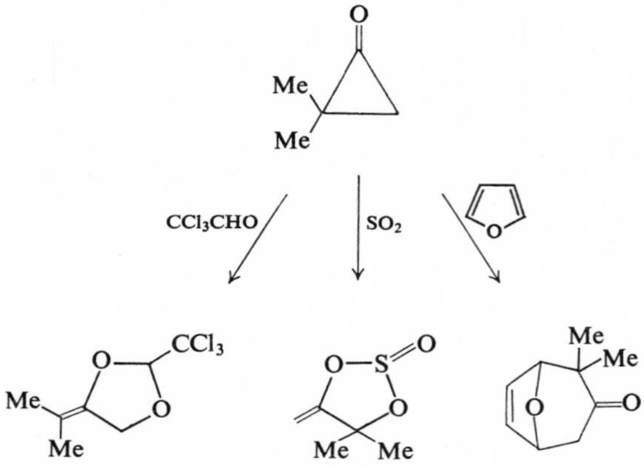

Fig. 5.19

There is no clear evidence that these additions are concerted; in particular a stepwise mechanism seems likely for the additions of trichloroacetaldehyde and sulphur dioxide.

Yet a further way in which $_\pi 4_s + _\pi 2_s$ addition could arise is in the addition of 4π cations to a monoene. The HOMO of the cation (63) is as shown in (64). Woodward has pointed out a reaction which can be fitted into this type of scheme (5.69).[8]

(63)

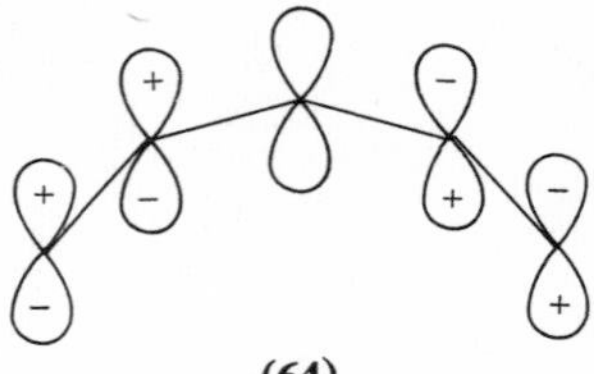

(64)

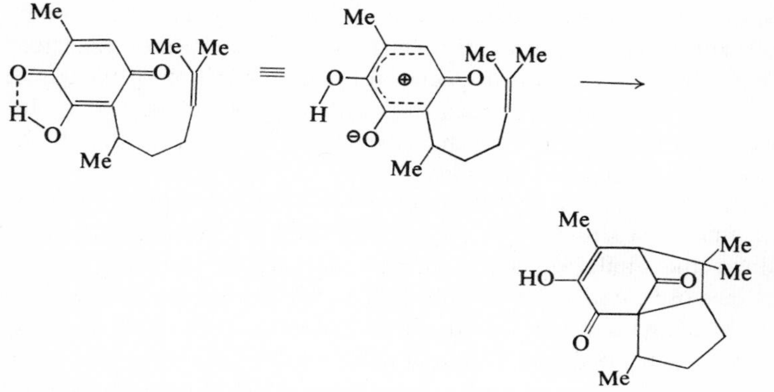

(5.69)

Photochemical cycloadditions.[59] Intermolecular cycloadditions involving six π electrons are rarely brought about photochemically. There is no firm evidence for the concerted ${}_{\pi}4_s + {}_{\pi}2_a$ or ${}_{\pi}4_a + {}_{\pi}2_s$ modes of addition which are predicted by the selection rules, assuming the simple photochemical excitation of an electron to the next highest orbital.

When mixtures of dienes or dienes and monoenes are irradiated some six-membered ring products are formed. However, these are usually minor products and are accompanied by considerable amounts of 2 + 2 adducts. The most reasonable explanation for this is that triplet excited

$$\underset{\text{sensitiser}}{\text{S}} \xrightarrow{h\nu} {}^{1}\text{S}^{*} \longrightarrow {}^{3}\text{S}^{*} \xrightarrow{\text{butadiene}} \text{S} + \underset{\textit{cisoid}\text{ and }\textit{transoid}}{{}^{3}(\text{butadiene})^{*}}$$

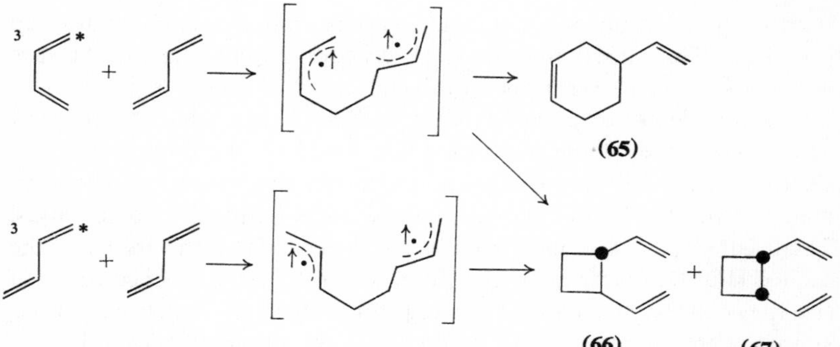

(65)

(66) (67)

Fig. 5.20

reactants are involved. These are formed directly (by sensitisers) or by intersystem crossing of initially formed singlets before intermolecular reaction. The triplet diradical species, formed by addition of the triplet excited reactant to an unexcited molecule, collapse mainly to cyclobutanes. A typical reaction scheme is shown in fig. 5.20, for the photosensitised dimerisation of butadiene.[60]

The ratio of cyclohexene [for example (65)] to cyclobutane derivatives [for example (66) and (67)] in the photosensitised dimerisation of butadiene and similar open chain dienes, depends on the energy of the triplet sensitiser. *cisoid* Diene molecules have a lower excitation energy than *transoid* and so can be selectively excited by lower energy sensitisers. The diradicals formed by addition of these *cisoid* triplet diene molecules have a greater chance of closing to cyclohexene than those from *transoid* dienes, which lead exclusively to cyclobutanes (see chapter 6).

Although intermolecular photochemical cycloadditions are unlikely to be concerted, intramolecular concerted cycloadditions are more likely since in these cases, initially formed excited singlet components can interact prior to relaxation to triplet states. *cis*-Hexatrienes are converted into bicyclo[3,1,0]hexenes on irradiation (fig. 5.21). This could involve concerted $_\pi 4_s + {}_\pi 2_a$ or $_\pi 4_a + {}_\pi 2_a$ processes but so far the reaction has not been carried out with sufficiently labelled reactants to clarify this.[8]

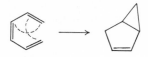

Fig. 5.21

In the irradiation of vitamin D_2, suprasterols I and II are formed (fig. 5.22). This clearly involves a $_\pi 2_a$ component but it cannot be seen whether the $_\pi 4$ component is suprafacial[8] as assumed in fig. 5.22.

The complexity of photochemical excitation, in all but the simplest cases, makes the simple selection rules, based on the total number of electrons involved, inapplicable to photochemical reactions. The symmetry of the HOMO of the excited species is frequently not obvious. The feasibility of the concerted mode also depends on the ability of the excited HOMO to interact with suitable orbitals on the other component (i.e. they must be close enough in energy for efficient mixing). For example, simple selection rules would predict that the photochemical $_\pi 4_s + {}_\pi 2_s$ addition of *o*-quinone to *trans*-stilbene (5.70) would be disallowed.

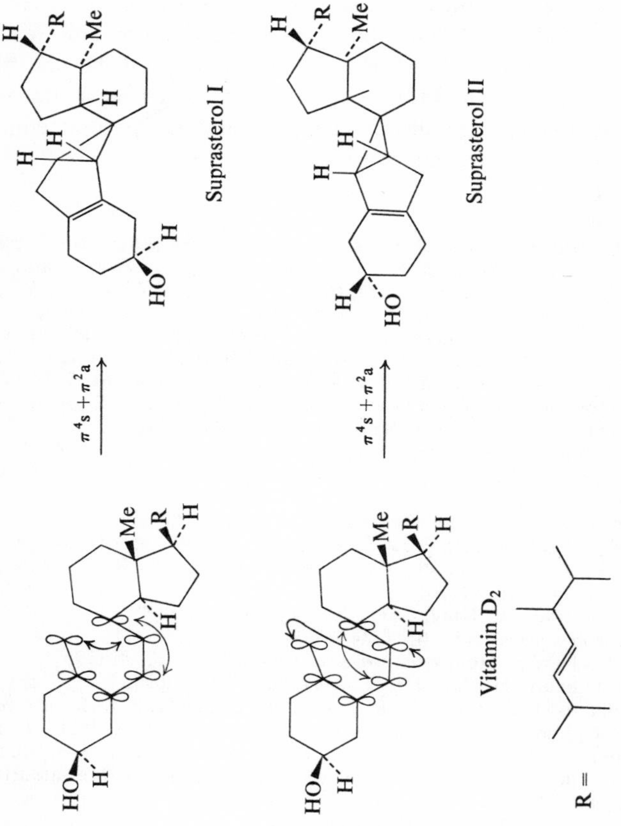

Fig. 5.22

$$\text{(structure)} + \text{(structure)} \longrightarrow \text{(structure)} \qquad (5.70)$$

However, calculations have shown that the stabilisation energy from interaction of appropriate vacant and occupied orbitals of an $n \to \pi^*$ excited o-quinone and ground state stilbene are of the same order of magnitude as those for a thermal Diels–Alder reaction.[61] This would suggest that the concerted photochemical addition would be feasible. In practice, the thermal addition of *trans*-stilbene to tetrachloro-o-benzoquinone is highly stereoselective and although the photochemical reaction is less so it is still remarkably stereoselective.[62]

REFERENCES

1. The scope and synthetic potential of the reaction have been the subject of many excellent reviews: (*a*) R. Huisgen, R. Grashey, and J. Sauer in *The Chemistry of Alkenes*, ed. S. Patai, Interscience, 1964; (*b*) J. Sauer, *Angew. Chem. Int. Edn*, **5**, 211 (1966); (*c*) A. Wassermann, *Diels–Alder Reactions*, Elsevier, 1965; (*d*) J. Hamer, *1,4-Cycloaddition Reactions*, Academic Press, 1967; and references in J. March, *Advanced Organic Chemistry*, McGraw-Hill, 1968. For a review of the stereochemistry, see J. G. Martin and R. K. Hill, *Chem. Reviews*, **61**, 537 (1961). For a review of the Diels–Alder reaction of heterodienes and dienophiles see S. B. Needleman and M. C. Chang Kuo, *Chem. Reviews*, **62**, 405 (1962).
2. R. W. Hoffmann, *Dehydrobenzene and Cycloalkynes*, Academic Press, 1967.
3. Reviews: (*a*) J. Sauer, *Angew. Chem. Int. Edn*, **6**, 16 (1967); (*b*) S. Seltzer, *Advances Alicyclic Chem.*, **2**, 1 (1968).
4. J. B. Lambert and J. D. Roberts, *Tetrahedron Letters*, 1965, 1457.
5. D. E. Van Sickle and J. O. Rodin, *J. Amer. Chem. Soc.*, **86**, 3091 (1964); W. R. Dolbier and S.-H. Dai, *ibid.*, **90**, 5028 (1968); P. Brown and R. C. Cookson, *Tetrahedron*, **21**, 1977, 1993 (1965).
6. Review: H. Kwart and K. King, *Chem. Reviews*, **68**, 415 (1968).
7. S. Seltzer, *J. Amer. Chem. Soc.*, **87**, 1534 (1965).
8. R. B. Woodward and R. Hoffmann, *Angew. Chem. Int. Edn*, **8**, 781 (1969).
9. W. C. Herndon and L. H. Hall, *Tetrahedron Letters*, 1967, 3095.
10. (*a*) K. L. Williamson, Y. F. L. Hsu, R. Lacko, and C. H. Youn, *J. Amer. Chem. Soc.*, **91**, 6129 (1969); K. N. Houk, *Tetrahedron Letters*, 1970, 2621; (*b*) for leading references see 10*a*, 1*c* and 3*a*. See L. Salem, *J. Amer. Chem. Soc.*, **90**, 553 (1968) for an alternative view of which secondary interactions are important in favouring *endo*-addition.
11. J. C. Little, *J. Amer. Chem. Soc.*, **87**, 4020 (1965).
12. R. B. Woodward and T. J. Katz, *Tetrahedron*, **5**, 70 (1959).
13. (*a*) R. Criegee and R. Askani, *Angew. Chem. Int. Edn*, **7**, 537 (1968); (*b*) see, however, H. W. Thompson and D. G. Melillo, *J. Amer. Chem. Soc.*, **92**, 3218 (1970) for a catalysed Diels–Alder reaction which appears to be stepwise.
14. K. Fukui in P.-O. Löwdin and B. Pullman, *Molecular Orbitals in Chemistry, Physics and Biology*, Academic Press, 1964.
15. (*a*) E. W. Garbisch and R. F. Sprecher, *J. Amer. Chem. Soc.*, **91**, 6785 (1969); (*b*) R. Sustmann, *Tetrahedron Letters*, 1971, 2721.

16. Review: R. Gompper, *Angew. Chem. Int. Edn*, **8**, 312 (1969).
17. I. Fleming and M. H. Karger, *J. Chem. Soc.* (C), 1967, 226.
18. E. B. Hershberg and J. R. Ruhoff, *Org. Synth.*, Coll. vol. II, 102.
19. R. B. Moffett, *Org. Synth.*, Coll. vol. IV, 238.
20. K. Alder, F. H. Flock, and H. Beumling, *Chem. Berichte*, **93**, 1896 (1960).
21. E. J. Corey and W. L. Mock, *J. Amer. Chem. Soc.*, **84**, 685 (1962).
22. E. Vogel, W. Grimme, and S. Korte, *Tetrahedron Letters*, 1965, 3625.
23. M. Avram, I. Dinulescu, M. Elian, M. Farcasiu, E. Marica, G. Mateescu, and C. D. Nenitzescu, *Chem. Berichte*, **97**, 372 (1964).
24. J. A. Berson and S. S. Olin, *J. Amer. Chem. Soc.*, **91**, 777 (1969).
25. M. J. Goldstein and G. L. Thayer, *J. Amer. Chem. Soc.*, **87**, 1925, 1933 (1965).
26. N. Rieber, J. Alberts, J. A. Lipsky, and D. M. Lemal, *J. Amer. Chem. Soc.*, **91**, 5668 (1969).
27. M. Martin and W. R. Roth, *Chem. Berichte*, **102**, 811 (1969).
28. E. L. Allred and J. C. Hinshaw, *Chem. Comm.*, 1969, 1021.
29. E. L. Allred, J. C. Hinshaw, and A. L. Johnson, *J. Amer. Chem. Soc.*, **91**, 3382 (1969).
30. Reviews (a) of the scope: R. Huisgen, *Angew. Chem. Int. Edn*, **2**, 565 (1963); (b) of the mechanism: *ibid.*, **2**, 633 (1963). The concept was first recognised by L. Z. Smith, *Chem. Reviews*, **23**, 193 (1938).
31. T. Kauffmann, H. Berg, and E. Koppelmann, *Angew. Chem. Int. Edn*, **9**, 380 (1970).
32. E. Buchner, *Berichte*, **21**, 2637 (1888); **23**, 701 (1890); A. Michael, *J. prakt. Chem.* (2), **48**, 94 (1893).
33. (a) W. F. Bayne and E. I. Snyder, *Tetrahedron Letters*, 1970, 2263; W. R. Dolbier and S.-H. Dai, *ibid.*, 4645; (b) R. Sustmann, *Tetrahedron Letters*, 1971, 2717.
34. S. Morrocchi, A. Ricca, and A. Zanarotti, *Tetrahedron Letters*, 1970, 3215.
35. S. Morrocchi, A. Ricca, A. Zanarotti, R. Gandolfi, G. Bianchi, and P. Gruenanger, *Tetrahedron Letters*, 1969, 3329.
36. R. A. Firestone, *J. Org. Chem.*, **33**, 2285 (1968).
37. R. Huisgen, *J. Org. Chem.*, **33**, 2291 (1968). It is very instructive to compare the arguments put forward by Firestone and by Huisgen in these two papers.
38. G. Wittig and W. Tochtermann, *Chem. Berichte*, **94**, 1692 (1961); F. Weygand and H. Daniel, *ibid.*, 1688.
39. R. Huisgen and H. Gotthardt, *Chem. Berichte*, **101**, 552, 839, 1059 (1968); R. Huisgen, H. Gotthardt, and R. Grashey, *ibid.*, **101**, 536, 829 (1968); H. Gotthardt, R. Huisgen, and R. Knorr, *ibid.*, **101**, 1056 (1968).
40. J. E. Baldwin and S. Y. Hong, *Chem. Comm.*, 1967, 1136; *Tetrahedron*, **24**, 3787 (1968).
41. (a) Review: B. P. Stark and A. J. Duke, *Extrusion Reactions*, Pergamon Press, 1967; (b) W. L. Mock and J. H. McCausland, *Tetrahedron Letters*, 1968, 391.
42. R. M. Dodson and R. F. Sauers, *Chem. Comm.*, 1967, 1189.
43. W. L. Mock, *J. Amer. Chem. Soc.*, **88**, 2857 (1966); S. D. McGregor and D. M. Lemal, *ibid.*, **88**, 2858 (1966).
44. J. Saltiel and L. Metts, *J. Amer. Chem. Soc.*, **89**, 2232 (1967).
45. D. M. Lemal and S. D. McGregor, *J. Amer. Chem. Soc.*, **88**, 1335 (1966); L. A. Carpino, *Chem. Comm.*, 1966, 494.
46. J. E. Baldwin, *Canad. J. Chem.*, **44**, 2051 (1966).
47. S. Yankelevich and B. Fuchs, *Tetrahedron Letters*, 1967, 4945, and references therein.
48. (a) A diradical mechanism has been demonstrated for the case of 7,7-dialkoxy-norbornadiene: R. W. Hoffmann and R. Hirsch, *Tetrahedron Letters*, 1970, 4819; (b) R. W. Hoffmann and H. Hauser, *Tetrahedron*, **21**, 891 (1965).

49. A. T. Blomquist and Y. C. Meinwald, *J. Amer. Chem. Soc.*, **81**, 667 (1959); J. K. Williams and R. E. Benson, *ibid.*, **84**, 1257 (1962).
50. M. Green, *J. Chem. Soc.*, 1965, 541.
51. C. D. Smith, *J. Amer. Chem. Soc.*, **88**, 4273 (1966).
52. N. Rieber, J. Alberts, J. M. Lipsky, and D. M. Lemal, *J. Amer. Chem. Soc.*, **91**, 5668 (1969).
53. M. Rosenblum, A. Longroy, M. Neveu, and C. Steel, *J. Amer. Chem. Soc.*, **87**, 5716 (1965).
54. D. H. R. Barton and B. J. Willis, *Chem. Comm.*, 1970, 1225.
55. (*a*) G. Wittig and R. W. Hoffmann, *Chem. Berichte*, **95**, 2718 (1962); (*b*) K. Sakai and J. P. Anselme, *Tetrahedron Letters*, 1970, 3851; (*c*) C. W. Rees and M. Yelland, personal communication.
56. J. F. M. Oth, *Angew. Chem. Int. Edn*, **7**, 646 (1968); *Rec. Trav. chim.*, **87**, 1185 (1968).
57. H. M. R. Hoffmann, D. R. Joy, and A. K. Suter, *J. Chem. Soc.* (B), 1968, 57; H. M. R. Hoffmann and D. R. Joy, *ibid.*, 1968, 1182.
58. S. S. Edelson and N. J. Turro, *J. Amer. Chem. Soc.*, **92**, 2770 (1970); N. J. Turro, *Accounts Chem. Research*, **2**, 25 (1969).
59. Reviews: A. A. Lamola and N. J. Turro, *Techniques of Organic Chemistry*, vol. xiv, ed. A. Weissberger, Interscience, 1969; O. L. Chapman and G. Lenz in *Organic Photochemistry*, vol. i, ed. O. L. Chapman, Arnold, 1967, 283.
60. R. S. H. Liu, N. J. Turro, and G. S. Hammond *J. Amer. Chem. Soc.*, **87**, 3406 (1965).
61. W. C. Herndon and W. B. Giles, *Chem. Comm.*, 1969, 497.
62. D. Bryce-Smith and A. Gilbert, *Chem. Comm.*, 1968, 1318, 1319.

6 Other cycloadditions

Most cycloadditions other than those with six electrons participating, involve four electrons. Considerations of orbital symmetry indicate that thermal $_\pi 2_s + {_\pi}2_s$ cycloaddition is not allowed but that concerted $_\pi 2_s + {_\pi}2_a$ addition (fig. 6.1) is.

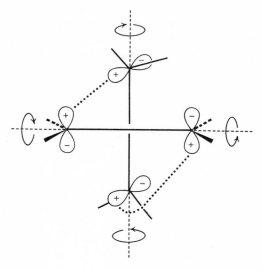

Fig. 6.1

For $_\pi 2_s + {_\pi}2_a$ addition the reactants must approach as shown in fig. 6.1. This involves inefficient orbital overlap and considerable twisting of the π bonds as the overlap is increased. Also the approach of the reactants is hindered by non-bonding steric interactions between substituents on the reactants. These effects combine to make the allowed $_\pi 2_s + {_\pi}2_a$ mode of addition very unfavourable so that $2\pi + 2\pi$ cycloadditions in

141

general proceed by stepwise mechanisms. The stepwise mechanisms are generally energy demanding so that $2\pi + 2\pi$ cycloadditions require forcing conditions unless the reactants bear substituents which are capable of stabilising either diradical or zwitterionic intermediates.

A variety of cumulated systems, for example allenes, undergo dimerisation or addition to π bonds to give four-membered rings by what are formally $2\pi + 2\pi$ cycloadditions. However, because the π bond orthogonal to the bond undergoing addition cannot be neglected, such $2\pi + 2\pi$ additions are considered separately.

6.1. Concerted $2\pi + 2\pi$ cycloadditions. No example of concerted thermal $_\pi 2_s + _\pi 2_a$ cycloaddition of two olefins has yet been unequivocably established. Because twisting of the π bonds of the reactants is necessary in such a cycloaddition the systems most likely to undergo $_\pi 2_s + _\pi 2_a$ addition are those in which the π bonds are initially twisted. Woodward and Hoffmann[1] have suggested that one of the products (2) obtained when benzene and butadiene are irradiated is the result of spontaneous concerted $_\pi 2_s + _\pi 2_a$ dimerisation of the initially formed photoadduct (1). The *trans* double bond in this photoadduct is necessarily twisted because it is constrained in an eight-membered ring. The Diels–Alder adduct of (1) and butadiene is also formed (6.1).[2]

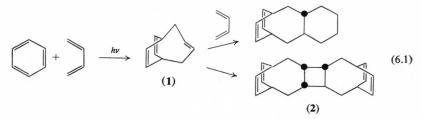

(1)

(2)

(6.1)

Fig. 6.2

cis,trans-Cyclo-octa-1,5-diene (3) also dimerises spontaneously at room temperature but the stereochemistry of the product is not known;[3] it would be predicted to be as shown in equation 6.2.

(3)   (4)   (6.2)

6.2. 2π + 2π Cycloadditions via ionic intermediates.[4] 2 + 2 Cyclo-additions through zwitterionic intermediates occur only when one component has strongly electron releasing groups and the other has strongly electron withdrawing groups, since such groups can consider-ably stabilise the positive and negative charge centres in the intermediate 1,4-dipole. This type of 2 + 2 addition often proceeds in high yield under very mild conditions. In some cases the additions are readily reversible, the dipolar intermediate being in equilibrium with both product and reactants. Orientation is always that resulting from the most stabilised zwitterion.

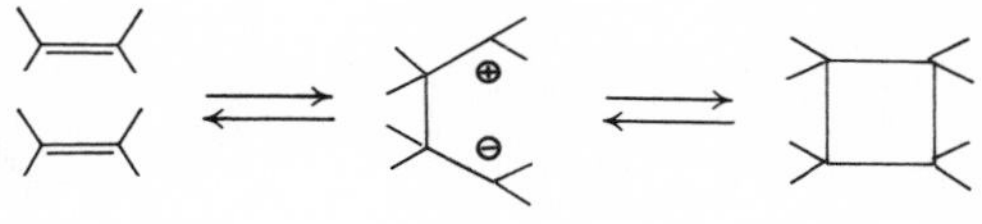

Typical olefins participating in ionic 2 + 2 additions are:

electron deficient component	*electron rich component*
$(CN)_2C{=}C(CN)_2$	$H_2C{=}CHOR$
$(CN)_2C{=}C(CN)Cl$	$H_2C{=}CHSR$
$(CF_3)_2C{=}C(CN)_2$	$H_2C{=}CHNR_2$
$CF_3(CN)C{=}C(CN)CF_3$	
$H_2C{=}CHCO_2R$	
$H_2C{=}CHSO_2R$	
$H_2C{=}CHNO_2$	

Some examples of these cycloadditions are shown in equations 6.3–6.5.

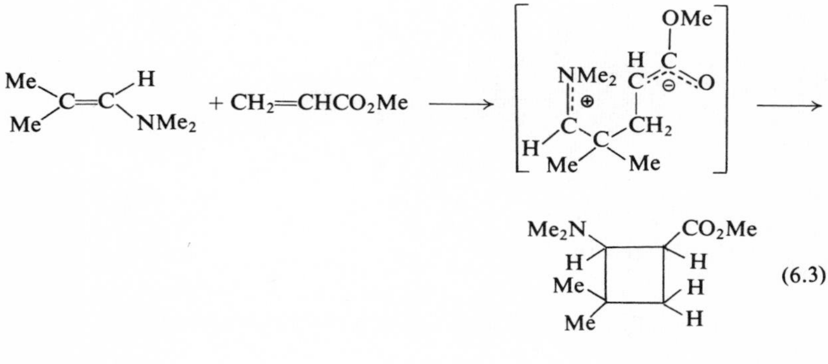

(6.3)

$$\text{MeO—} \bigcirc \text{—CH=CH}_2 + \underset{NC}{\overset{NC}{>}} C = C \underset{CN}{\overset{CN}{<}} \longrightarrow \qquad (6.4)$$

$$CH_2 = C \underset{S-CMe_3}{\overset{H}{<}} + \underset{NC}{\overset{F_3C}{>}} C = C \underset{CN}{\overset{CF_3}{<}} \longrightarrow +$$

$$\qquad\qquad\qquad\qquad\qquad\qquad\qquad\qquad (6.5)$$

On mixing, these olefins form deeply coloured solutions. The colours, which are attributed to π complexes formed between the electron rich and electron deficient olefins, are discharged as the reaction proceeds.

The characteristics of these reactions are consistent with a stepwise process involving a zwitterionic intermediate. The rates frequently show great dependence on solvent polarity, being much greater in more polar solvents; for example the addition shown in equation 6.4 is 6.3×10^4

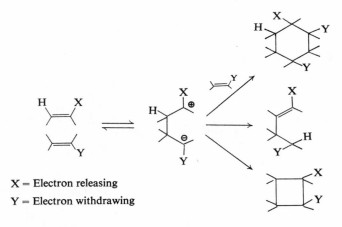

X = Electron releasing
Y = Electron withdrawing

Fig. 6.3. Reaction of electron rich with electron deficient olefins.

times faster in acetonitrile than in cyclohexane.[4a,5] In some cases, however, the additions are only moderately sensitive to solvent polarity (see p. 16).

The additions are extremely sensitive to substituent effects. For example, addition of styrenes to tetracyanoethylene only occurs when the styrenes have electron releasing *para*-substituents [e.g. *p*-OMe, equation 6.4] and the rate of these additions varies greatly with the electron releasing ability of the *para*-substituent: $k_{p\text{-OMe}}/k_{p\text{-cyclopropyl}} = 500.$[4d]

Further strong evidence for the stepwise nature of these cycloadditions comes from the observation of alternative reactions of the 1,4-dipolar intermediates. These are shown schematically in fig. 6.3.

Proton transfer can compete with ring closure when the electron rich component has a β-hydrogen atom. For example, the enamine (**5**) gives cyclobutane (**6**) with nitro-olefins in non-polar solvents but gives nitroenamines (**7**) in more polar solvents (6.6). It is assumed that the intermediate zwitterion is better stabilised in the more polar solvents and is therefore more selective so that proton transfer competes with ring closure.[6] Such competing reactions are general for enamines with β-hydrogen atoms.

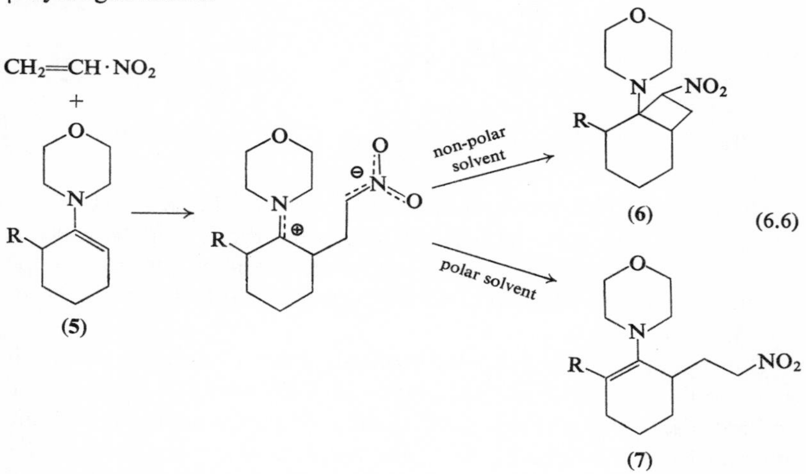

$$CH_2{=}CH{\cdot}NO_2$$

(6.6)

(**5**)   (**6**)   (**7**)

Reaction of the 1,4-dipole with a second molecule of electron deficient reactant to give a six-membered ring 1:2-adduct may also occur. The enamine (**8**) and β-nitrostyrene normally give the cyclobutane but in the presence of excess β-nitrostyrene the 1:2-adduct (**9**) is isolated (6.7).[6]

$$PhCH{=}CH\cdot NO_2$$
$$+$$
$$EtCH{=}CH{-}N\bigcirc$$
(8)

$\longrightarrow$

(1,4-dipole intermediate: Et, Ph substituents, $\overset{\oplus}{N}$-pyrrolidine, $\overset{\ominus}{}$H, NO_2)

$\overset{PhCH{=}CHNO_2}{\nearrow}$

cyclobutane adduct (Et, H, H, Ph, H, H, NO_2, N-pyrrolidine)

six-membered adduct **(9)** (Et, H, Ph, H, NO_2, N-pyrrolidine, H, NO_2, Ph, H)

(6.7)

$$CH_2{=}C\overset{CO_2Et}{\underset{CO_2Et}{}}$$
$$+$$
$$\overset{Me}{\underset{Me}{}}C{=}C\overset{H}{\underset{NMe_2}{}}$$
(10)

$\longrightarrow$

1,4-dipole intermediate **(with Me, Me, $\overset{\oplus}{N}$–Me, Me, $\overset{\ominus}{C}$, CO_2Et groups)**

$\longrightarrow$

six-membered ring adduct **(11)** (Me, Me, Me_2N, H, EtO_2C, CO_2Et, CO_2Et, H, H, CO_2Et)

(6.8)

The enamine **(10)** forms cyclobutanes with acrylic acid derivatives (equation 6.3) but with methylenemalonic ester, the 1,4-dipole, which is even better stabilised, reacts with another molecule of ester to give the six-membered ring adduct **(11)** rather than closing to form the cyclobutane (6.8).[7]

This type of six-membered ring formation from a 1,4-dipole and a 1,4-dipolarophile is known as 1,4-dipolar cycloaddition.[8] It is quite general and promises great synthetic application. 1,4-Dipoles are fundamentally different from 1,3-dipoles(§ 5.3). The latter have 4 π electron systems extending over the three atoms and the formal terminal positive and negative charges are interchangeable. In 1,4-dipoles two saturated atoms separate the charges which are not interchangeable.

In certain cases 1,4-dipolar intermediates can actually be isolated, for example, the exceptionally well stabilised dipoles **(14)** in equation 6.9.[4a]

The yellow to orange crystalline 1,4-dipoles are formed at low temperature but at higher temperatures they close to cyclobutanes which then undergo elimination to give cyclobutenes. Electrocyclic ring opening of the cyclobutenes gives butadienes, the observed products. The 1,4-dipoles from (**12**, X = SMe) and *p*-nitrobenzylidene malononitrile (**13**, $R^4 = p\text{-}NO_2C_6H_4$, $R^5 = H$) or 1,1-dicyano-2,2-bistrifluoromethyl ethylene (**13**, $R^4 = R^5 = CF_3$) are particularly stable and only rearrange to butadienes on heating to about 80°C.

$$\text{(6.9)}$$

X = SMe, NMe_2 $R^2 + R^3 = (CH_2)_{2-4}$
R^1 = H, Me, Et, Pr^i,   Ph R^5 = H, CF_3, CN
R^4 = Ph, $p\text{-}NO_2C_6H_4$, CF_3, CN

Lack of stereospecificity has been observed in ionic 2 + 2 cycloadditions, in line with the presence of an intermediate in which bond rotation can occur before closure of the second bond.[9,4a,4d] In general, however, ionic additions are highly stereoselective; this contrasts with the much lower stereoselectivity observed for cycloadditions with diradical intermediates. Ionic 2 + 2 additions are much more stereoselective because the electrostatic attraction between the ends of the 1,4-dipole favours its formation in a configuration approaching (**15**); the second bond may therefore be formed without rotation of other σ bonds. In the diradical intermediate no such electrostatic attraction is present and steric factors tend to make an open configuration (**16**) more likely. This is particularly so since the carbon atoms bearing the unpaired electrons tend to have the bulky substituents which stabilise the radical centres.[4d] In the time taken for bond rotation to bring the diradical into a suitable conformation for ring closure, greater loss of configuration can occur.

(15)   **(16)**

As the solvent polarity is increased, open 1,4-dipolar configurations, corresponding to **(16)**, are likely to have greater stability so that selectivity should fall. This is observed in the addition of *cis*-anethole ($MeOC_6H_4CH{=}CHMe$) to tetracyanoethylene which gives 10–15 per cent *trans*-cyclobutane in benzene and 49 per cent in acetonitrile. With *trans*-anethole only *trans*-cyclobutane is formed even in the most polar solvent, showing that there is a greater driving force for rotation about the appropriate σ bond in the intermediate formed from *cis*-anethole which initially has the bulky aryl and methyl groups *cis*.[4d]

The very high stereoselectivity of some of these cycloadditions raises the question of whether they should be considered as concerted. If the charge transfer interaction between the ends of the dipole (which ultimately leads to stereospecific bond formation) develops as the charges develop in the dipole, the process is effectively multicentre. The problem of deciding whether the energy profile has a single peak or a dip, corresponding to an intermediate in which the charge centres are bound, as in an intimate ion pair, is as yet insoluble. It is probable that there is a spectrum of mechanisms ranging from the completely stepwise with a discrete intermediate zwitterion to an asymmetric multicentre process. This type of 'two-stage' concerted reaction is, however, basically different from an orbital symmetry-allowed concerted process.

Four-membered ring formation by an ionic cycloaddition is not limited to cyclobutanes. Imines dimerise to give 1,3-diazetidines (6.10). These are frequently unstable and revert to imines which may or may not be the same as the starting imines.[4c]

Azodicarboxylic ester, in addition to being a reactive dienophile, also undergoes $2+2$ additions with electron rich olefins to give 1,2-diazetidines (6.11).[10,11a]

$$R^1—CH=N—R^2 + R^3—N=CH—R^4 \rightleftharpoons \begin{array}{c} R^1 \\ CH—N \\ \nearrow \qquad \qquad \searrow \\ R^2 \\ N—CH \\ R^3 \qquad R^4 \end{array} \rightleftharpoons \begin{array}{c} R^1 \qquad R^2 \\ CH \qquad N \\ \| \qquad \| \\ N \qquad CH \\ R^3 \qquad R^4 \end{array} \qquad (6.10)$$

$$\begin{array}{c} CH_2=C \overset{OR}{\underset{H}{<}} \\ \dfrac{k_H}{k_D} = 0.83 \qquad \dfrac{k_H}{k_D} = 1.12 \\ RO_2C—N=N—CO_2R \end{array} \longrightarrow \begin{array}{c} CH_2—C \overset{OR}{\underset{H}{<}} \\ | \qquad | \\ N—N \\ RO_2C \qquad CO_2R \end{array} \qquad (6.11)$$

With vinyl ethers, the addition is stereospecific and almost solvent-independent; the values of $\Delta H^{\ddagger}$ and $\Delta S^{\ddagger}$ also support a concerted addition. The secondary H/D isotope effects are shown. They are remarkably similar to those observed for the concerted addition of keten to styrene and suggest that the two-bond forming processes occur simultaneously but are very different at each terminus. The workers prefer to explain these different isotope effects by a stepwise reaction through a dipolar intermediate, an idea supported by the trapping of a dipolar intermediate in the $2+2$ addition of indene to 4-phenyl-1,2,4-triazoline-3,5-dione.[11a] Experimental observations such as these underline the need for caution in deciding on concerted or stepwise pathways. Obviously further subtle investigation of this formally disallowed, but apparently concerted, $2+2$ cycloaddition is required. With symmetrical, less electron rich olefins, azodicarboxylic ester acts as a diene to give formal Diels–Alder adducts (6.12).[11a,11b]

$$\begin{array}{c} O \diagdown \quad OR \\ \| \\ N^{\diagdown} N \\ | \\ CO_2R \end{array} \longrightarrow \begin{array}{c} O \diagdown \quad OR \\ \\ N—N \\ | \\ CO_2R \end{array} \qquad (6.12)$$

Many other cycloadditions are known in which hetero-1,4-dipoles are formed as intermediates.[4a] These give four-membered ring heterocycles by ring closure or six-membered ring heterocycles by addition to another component, as, for example, in equation 6.13.[12] Many of these involve cycloadditions of heterocumulenes, which are discussed later.

$$\text{(6.13)}$$

6.3. $2\pi + 2\pi$ Cycloadditions via radical intermediates.[4b,4d,13] Ionic additions occur when the components bear substituents capable of stabilising a zwitterionic intermediate. This is therefore when the electronic demands of the substituents are complementary. Other alkenes or alkynes which have activating groups capable of stabilising a diradical intermediate, also undergo 2 + 2 additions, but forcing conditions are generally required. Typically these cycloadditions are carried out at 100–200°C under pressure.

Polychlorofluoro-olefins are exceptionally reactive in this type of addition. They dimerise or react with other activated olefins. Also striking is their often almost exclusive 2 + 2 addition to dienes; for example, dichlorodifluoroethylene reacts with butadiene to give a vinyl cyclobutane (equation 6.17), whereas ethylene and most other olefins give Diels–Alder adducts.

Evidence that these are radical additions comes from the observed orientations, which are always those which would arise from the most stable diradical intermediate. Thus the dimerisation of acrylonitrile (6.14) leads to 1,2-dicyanocyclobutane since the diradical intermediate (**17**), in which both radical centres are stabilised by α-cyano-groups, is more stable than the alternative diradical (**18**). This orientation is not consistent with an ionic mechanism since the zwitterion (**19**), in which the positive charge would be destabilised by the strongly electron withdrawing cyanide, should be less stable than the zwitterion (**20**) which would give 1,3-dicyanocyclobutane.[13a] Other examples which illustrate the orientation are shown in equations 6.15–6.17. An α-chlorine is better able to stabilise a radical centre than an α-fluorine; thus the

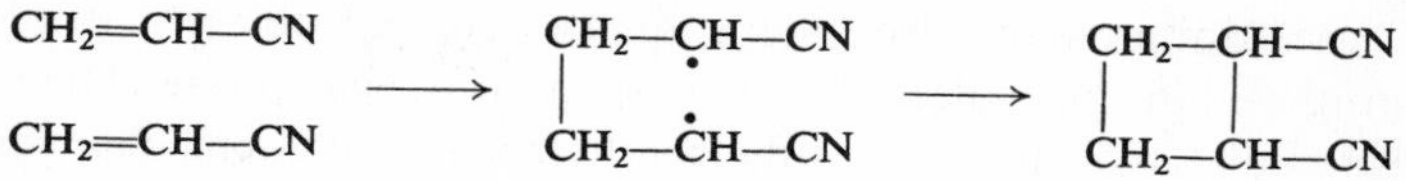

CH₂=CH—CN

CH₂=CH—CN

⟶

CH₂—CH—CN (•)

CH₂—CH—CN (•)

(**17**)

⟶

CH₂—CH—CN

CH₂—CH—CN

(6.14)

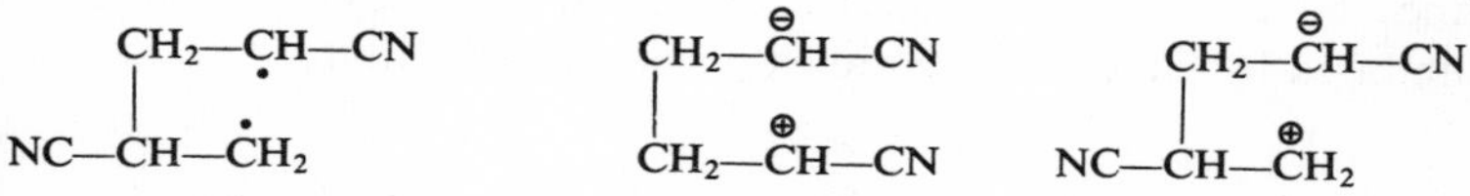

CH₂—CH—CN (•)

NC—CH—CH₂

(**18**)

CH₂—$\overset{\ominus}{\text{CH}}$—CN

CH₂—$\overset{\oplus}{\text{CH}}$—CN

(**19**)

$\overset{\ominus}{\text{CH}}$₂—CH—CN

NC—CH—$\overset{\oplus}{\text{CH}}$₂

(**20**)

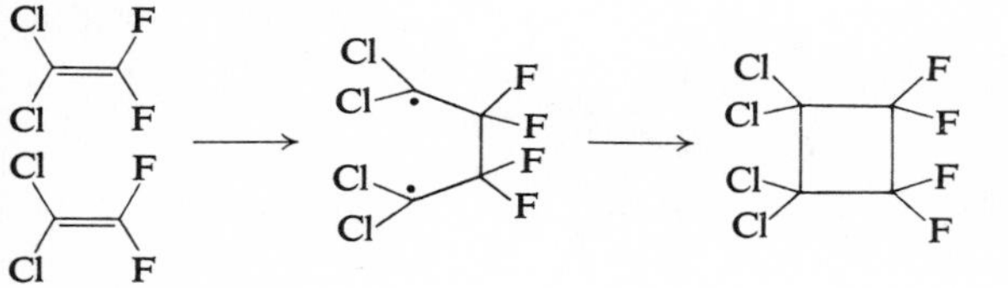

(6.15)

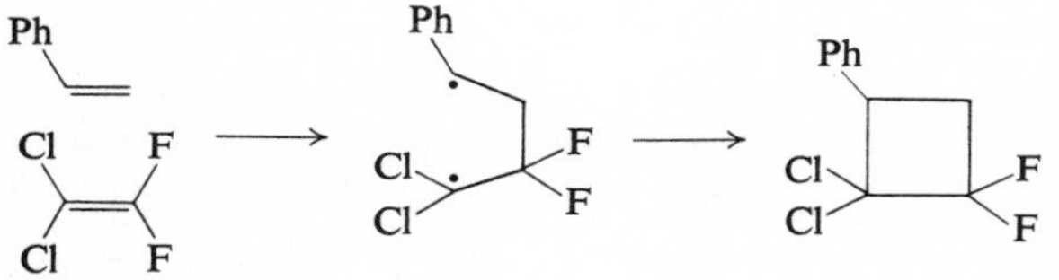

(6.16)

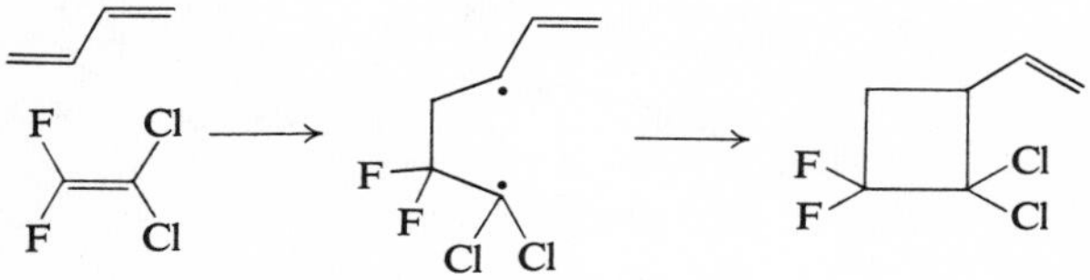

(**21**)

(6.17)

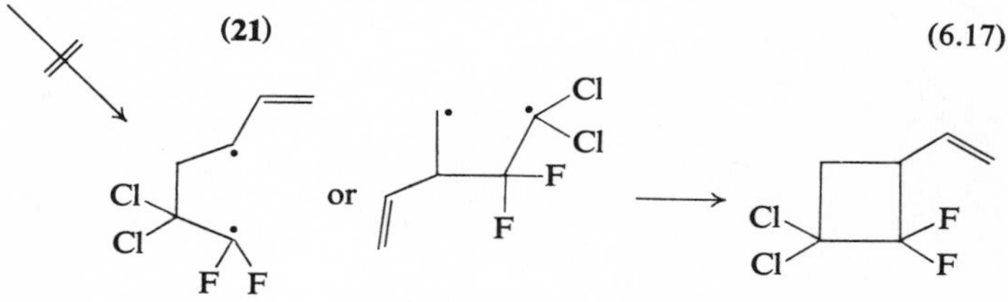

(**22**)   or   (**23**)

addition of dichlorodifluoroethylene to butadiene proceeds as shown in equation 6.17. The diradical (**21**) is estimated to be approximately 8 kcal mol⁻¹ (33 kJ mol⁻¹) lower in energy than (**22**) and more than 21 kcal mol⁻¹ (88 kJ mol⁻¹) lower than (**23**). Such energy differences are sufficient to ensure that the orientation is virtually all as shown.[13b]

These additions are also independent of solvent polarity, often proceeding well in non-polar solvents or even in the gas phase. The rate of addition of 1,1-dichloro-2,2-difluoroethylene to butadiene changes by a factor of less than three in going from hexane to methanol as solvent.[4d]

Since the reactions are not radical chain reactions they are not initiated or inhibited by the usual free radical initiators or inhibitors.

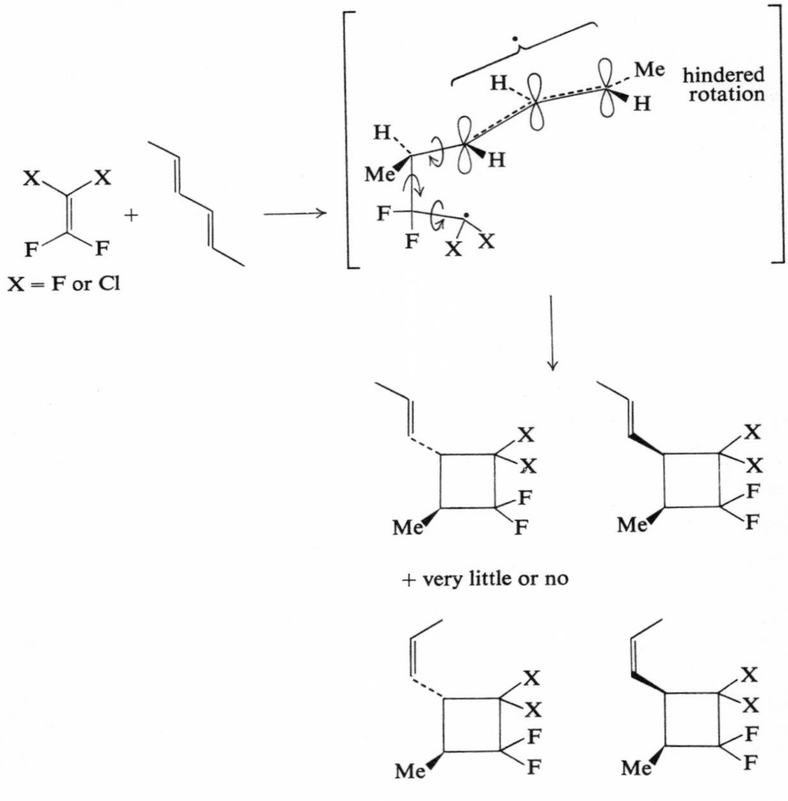

Fig. 6.4

Radical 2 + 2 additions show the lack of stereospecificity expected from an intermediate in which free rotation about σ bonds can occur. The diradical intermediate for a thermal addition should be formed as a singlet so that there is no spin inversion barrier to ring closure. However, the conformation in which the diradical is initially formed is unlikely to be right for immediate ring closure. Thus, bond rotations can occur in the time taken for the molecule to attain the required conformation for closure of the second bond. For example, in the addition of 1,1-dichloro-2,2-difluoroethylene and tetrafluoroethylene to *trans, trans*-hexa-2,4-diene and the isomers of 1,4-dichlorobutadiene, loss of configuration occurs at the double bond which becomes part of the cyclobutane ring, but, as expected, configuration at the other double bond is largely or completely retained because of the barrier to rotation in the allyl radical residue (fig. 6.4).[14,4d]

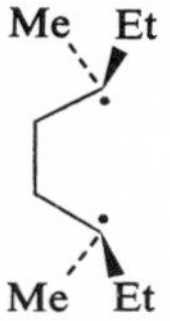

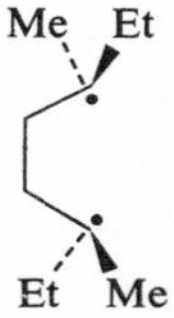

The intermediate diradicals can revert to reactants rather than cyclise. This is shown by the recovery of isomerised dienes from the above reaction. Since the dienes do not themselves isomerise under the reaction conditions, the isomerised dienes must result from cleavage of the diradicals after rotation has occurred.[15]

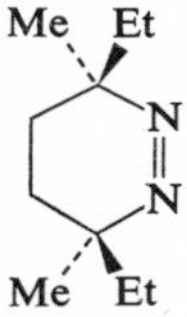

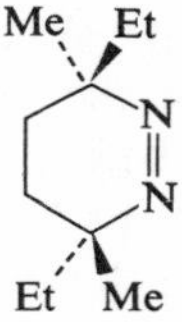

Diradicals are also formed in photochemical cycloadditions, but here they are often triplet diradicals. In this case, in addition to the diradical having to attain the right conformation, there is a further barrier to ring closure, namely spin inversion. It is not clear how the rate of spin inversion in this type of triplet diradical compares with the rate of bond rotation. There is some experimental evidence to suggest that triplet diradicals have a longer lifetime prior to ring closure than the singlet diradicals produced in the thermal reaction. There is a greater loss of configuration in photosensitised triplet diradical additions than in the thermal reaction.[4d] Also when the diradicals (24) and (25) are generated in the singlet state by thermal decomposition or by direct photolysis of the azo-compounds (26) and (27), they show a different pattern of products from the corresponding triplet diradicals generated by photo-sensitised decomposition.[4d]

Both types of diradical are formed in conformations which are ideally disposed for ring closure, but cleavage to methylbut-1-ene represents a large proportion of the reaction. Significantly, cyclobutane formed from the singlet diradical almost completely retains the con-figuration of the azo compound, but the triplet diradical gives cyclo-butane with appreciable loss of configuration. A greater proportion of cleavage also occurs for the triplet diradical, in line with it having a greater barrier to ring closure.

A possible alternative explanation of these stereochemical results is based on the conformational flexibility of the intermediate rather than on spin. Two ground state olefins cannot correlate directly with ground state cyclobutane in any geometrically accessible way: as the first bond is formed, there is a symmetry-imposed barrier to concerted formation of the second bond. An intermediate is produced which has at least a conformational barrier to ring closure. This species, identifiable with the 'singlet' of the previous treatment, can therefore undergo loss of configuration before closure: the cycloaddition may þe stereoselective, but is unlikely to be stereospecific. An olefin in an excited electronic configuration will react with a ground state olefin to produce an excited configuration of the intermediate; this may well have different barriers to conformational change – it may be more 'floppy' – and may therefore close less stereoselectively than the ground state species. This type of argument, backed up by molecular orbital calculations, has been applied successfully to the similar additions of singlet and triplet carbenes (see § 6.6).

Concerted versus stepwise radical additions to dienes.[4d,13b] Generally the concerted $4 + 2$ Diels–Alder reaction is the most favourable mode of addition of a monoene to a diene. However, if this mode of addition is particularly disfavoured or if formation of a diradical is particularly favoured, a diradical addition can compete. In some cases the diradical formation may be sufficiently favourable compared with the concerted Diels–Alder reaction for it to occur almost to the exclusion of the latter. The addition of 1,1-dichloro-2,2-difluoroethylene to butadiene is just such an example. It can be seen from table 6.1 that the addition gives mainly cyclobutane but a trace of cyclohexene is also formed. This is to be expected from the diradical mechanism. The butadiene exists in *cisoid* or *transoid* conformations which should have essentially the same

TABLE 6.1. *Addition of olefins to butadiene*

Olefin	Percentage composition of product	
	2 + 2 adduct	2 + 4 adduct
$CF_2{=}CCl_2$	99	1
$CHF{=}CF_2$	85.8	14.2
$CH_2{=}CF_2$	35	65
$CH_2{=}C{<}^{OAc}_{CN}$	14	86
$CH_2{=}CH_2$	0.02	99.98

reactivity for diradical formation. The diradical is allyl stabilised and therefore bond rotations within the allyl residue will be hindered compared with rotations about the other bonds. The diradical formed from a *transoid* diene cannot close to a cyclohexene without rotations within this allyl portion of the molecule. Such diradicals would therefore be expected to close exclusively to cyclobutanes. On the other hand, diradicals formed from *cisoid* butadiene can close to both cyclobutane or cyclohexene without rotations within the allyl system. Since butadiene and most other open chain dienes exist predominantly in the *transoid* form, the small amount of cyclohexene produced can be explained solely in terms of diradical formation (fig. 6.5).

It can be seen that cyclobutane formation is favoured over cyclohexene formation even for the *cisoid* diene since conformations of diradical which lead to the cyclohexenes are less favourable than those which lead to cyclobutanes.

In this type of radical addition, where rotation within the allylic system is disfavoured because of delocalisation, the proportion of cyclohexene should parallel, but never exceed, the proportion of the diene which is in a *cisoid* conformation (provided of course, that the rates of diradical formation from *cisoid* and *transoid* dienes are comparable). This is found to be the case. There is a close correlation between the amount of 2 + 4 adduct of 1,1-dichloro-2,2-difluoroethylene to butadiene and the fraction of butadiene in a *cisoid* conformation as this is varied by changing the temperature.[16] For 2-substituted butadienes, the proportion of 2 + 4 adduct increases with the size of the 2-substituent, since the larger the 2-substituent the greater the proportion of *cisoid* diene.[17]

Although the small amount of cyclohexene derivative formed in the addition of 1,1-dichloro-2,2-difluoroethylene to butadiene can be accommodated by the diradical mechanism, it is also possible that some or all of this 2 + 4 adduct arises by a competing Diels–Alder reaction which would be equally favoured by an increasing fraction of *cisoid* diene.

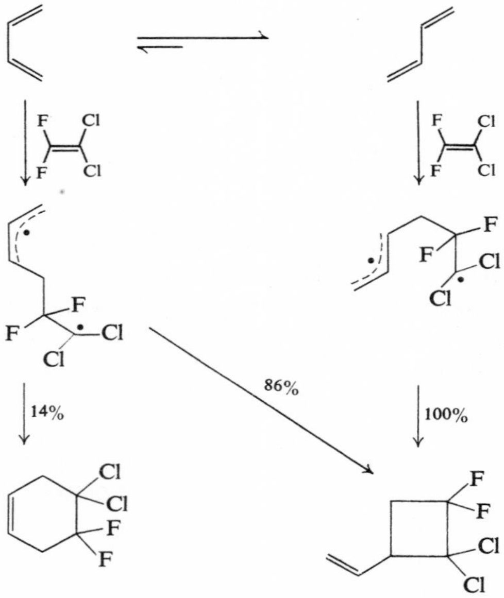

Fig. 6.5

Where the proportion of 2 + 4 adduct rises above the proportion of *cisoid* diene, it seems certain that the 2 + 4 adducts arise largely by the concerted Diels–Alder mechanism and that the 2 + 2 adducts arise by a competing stepwise reaction. For example, the thermal addition of trifluoroethylene to butadiene at 215°C gives 85.8 per cent 2 + 2 and 14.2 per cent 2 + 4 adducts.[4d] Taking the photosensitised addition of trifluoroethylene to butadiene as a model for the reaction proceeding entirely through a diradical intermediate, it can be estimated that only 2 per cent of 2 + 4 adduct in the thermal reaction would arise from the radical mechanism. The remainder therefore results from a concerted Diels–Alder reaction. The additions of 1,1-difluoroethylene and trifluoroethylene (table 6.1) are therefore examples where diradical formation and concerted Diels–Alder reaction are energetically comparable. This is reasonable because fluorine is less able to stabilise a radical than chlorine, so these olefins would give less stable diradicals than that from 1,1-dichloro-2,2-difluoroethylene, which reacts almost entirely by the diradical mechanism. As expected, diradical formation by trifluoroethylene is less favourable than concerted Diels–Alder reaction when *cis* fusion of the diene favours the latter. Thus the thermal addition of trifluoroethylene to cyclopentadiene gives almost exclusively 2 + 4 adduct with less than 0.1 per cent 2 + 2 adduct. When the addition is forced to proceed by the stepwise radical mechanism, as in the photosensitised reaction, the product contains 87 per cent 2 + 2 and 13 per cent 2 + 4 adduct.[4d].

In principle, a simple way to determine whether 2 + 2 or 2 + 4 adducts arise by competing stepwise and concerted modes would be to study the stereoselectivity of the reaction. In practice there are few systems which give both 2 + 2 and 2 + 4 adducts simultaneously where this can be studied. The only example reported so far is the addition of *cis*- and *trans*-1,2-dichloro-1,2-difluoroethylene to cyclopentadiene. This gives more than 95 per cent of 2 + 4 adduct stereospecifically. The minor portion of 2 + 2 adduct is formed non-stereospecifically, in line with concerted 2 + 4 and stepwise 2 + 2 mechanisms.[18]

α-Acetoxyacrylonitrile gives, with butadiene, a ratio of 2 + 2 to 2 + 4 addition that is little affected by variation of solvent or temperature. This was originally taken as evidence that both types of adduct were formed through a common diradical intermediate, since it seemed unlikely that competing mechanisms would depend so similarly on conditions.[19] This is now better interpreted as a competition between concerted 2 + 4 and stepwise 2 + 2 additions, because the related addition of *trans,trans*-hexadiene to α-acetoxyacrylonitrile gives only 2 + 4 adduct, stereospecifically and therefore almost certainly by a concerted mechanism.[20] Introduction of the two electron-releasing methyl groups into butadiene favours the Diels–Alder reaction (p. 94) and hinders radical addition (initial attack on a secondary carbon is less favourable than on a primary carbon) so that any competing radical process falls below the level of detection.

Careful investigation of the addition of ethylene to butadiene discloses a trace of vinylcyclobutane.[21] This can be accounted for by a radical reaction which competes unfavourably with the concerted Diels–Alder process. It is obvious that a negligible amount of the 2 + 4 adduct can be formed by the radical route.

6.4. 2 + 2 Cycloadditions of cumulenes.[22,46]   Cumulenes have at least two orthogonal π bonds, with a central atom common to both of the π bonds. Allene (**28**) is a cumulene with an all-carbon skeleton. Hetero-cumulenes can be considered as being derived from allene by replacement of one, two or three carbon atoms by hetero-atoms; keten (**29**) is an example.

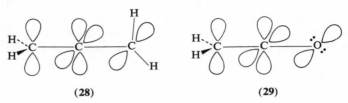

(**28**)   (**29**)

Table 6.2 shows the types of systems which are known. Some 1,3-dipolar species can be written with cumulenic resonance forms, for example

$$R\text{—}\overset{\oplus}{C}\text{≡}\overset{\ominus}{N}\text{—}O \leftrightarrow R\text{—}\overset{\ominus}{C}\text{=}\overset{\oplus}{N}\text{=}O$$

but only those systems where the main uncharged resonance form has a cumulene structure will be considered.

The reactivity and stability of heterocumulenes varies considerably. Sulphur dioxide, carbon disulphide and carbon dioxide are very stable, relatively unreactive compounds. Ketens are of intermediate reactivity; some diaryl ketens, though reactive, are isolable, but other ketens have to be generated *in situ*. At the other extreme, sulphenes have so far only been detected and trapped as reactive intermediates.

The separate π bonds in cumulenes can act as dienophiles or dipolaro-philes in Diels–Alder or 1,3-dipolar cycloadditions, and molecules in which a double bond is conjugated with the cumulene can function as dienes in Diels–Alder additions (fig. 6.6). In general, however, the notable and characteristic feature of the cycloaddition reactions of cumulenes is the tendency to form four-membered rings. Such reactions are formally $2\pi + 2\pi$ cycloadditions, and were therefore originally

Fig. 6.6

thought to go by stepwise mechanisms. This seemed reasonable because the sort of intermediates involved would, in general, be well stabilised zwitterions or allyl diradicals. This general picture had to be re-examined, however, when the cycloadditions of ketens to certain olefins were shown to be concerted.

Heterocumulene cycloadditions. Mechanism. Since concerted thermal $_\pi 2_s + _\pi 2_s$ cycloadditions are not allowed, considerable interest was

TABLE 6.2. *Heterocumulenes*

$R_2C{=}C{=}O$	ketens	
$O{=}C{=}C{=}C{=}O$	carbon suboxide	
$R_2C{=}C{=}NR$	ketenimines	one hetero-atom
$R_2C{=}C{=}S$	thioketens	
$R_2C{=}S{=}CR_2$	sulphilidines	
$O{=}C{=}O$	carbon dioxide	
$S{=}C{=}S$	carbon disulphide	
$RN{=}C{=}NR$	carbodi-imides	
$R_2C{=}S{=}O$	sulphines	
$R_2C{\overset{\overset{\displaystyle O}{\|}}{=}}S{=}O$	sulphenes	two hetero-atoms
$S{=}C{=}O$	carbonyl sulphide	
$RN{=}C{=}O$	isocyanates	
$RN{=}C{=}S$	isothiocyanates	
$O{=}N{=}O$	nitrogen dioxide	
$O{=}S{=}O$	sulphur dioxide	
$RN{=}S{=}NR$	sulphur di-imines	three hetero-atoms
$RN{=}S{=}O$	sulphinyl amines	
$RN{\overset{\overset{\displaystyle O}{\|}}{=}}S{=}O$	sulphuryl amines	

aroused by the stereospecific additions of olefins to ketens.[23] It seemed even more remarkable that ketens form only the disallowed 2 + 2 adduct, and none of the allowed 2 + 4 adduct, with dienes.[23,24]

Huisgen[25] closely investigated the cycloadditions of olefins to ketens and concluded that the reaction was in fact concerted but that bond formation in the transition state was unequal. The additions are highly stereospecific, with low ΔH^+ ($\sim$9–10 kcal mol^{-1}; 38–42 kJ mol^{-1}) and large negative ΔS^+ (~ -40 e.u.; 1 e.u. = 4.18 J mol^{-1} K^{-1}). The effect of solvent polarity on the rate of reaction is small but consistent with some charge build-up in the transition state. Thus, for example, a rate difference of about 50 is observed in the changing from cyclohexane to acetonitrile, compared with 63 000 for the stepwise ionic addition of tetracyanoethylene to *p*-methoxystyrene. This small charge build-up in the transition state explains the orientation of the additions which is such that substituents are best able to stabilise the developing charges (fig. 6.7).

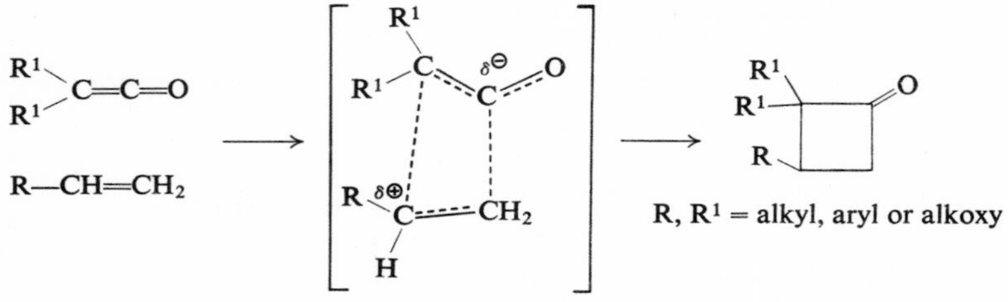

Fig. 6.7

The addition of ketens to norbornene proceeds without any of the skeletal rearrangement which might be expected from a stepwise reaction involving a zwitterion with a norbornyl cation component. Finally, Baldwin has shown that for the addition of styrene to diphenylketen, secondary deuterium isotope effects occur at each terminus. The magnitudes of these effects are interesting (fig. 6.8): that for bond *a* is consistent with a change of hybridisation from sp^2 to sp^3 in the rate determining step. However, that for bond *b* is not consistent with a simple change in hybridisation but does show that some form of interaction occurs in the rate determining step.[26]

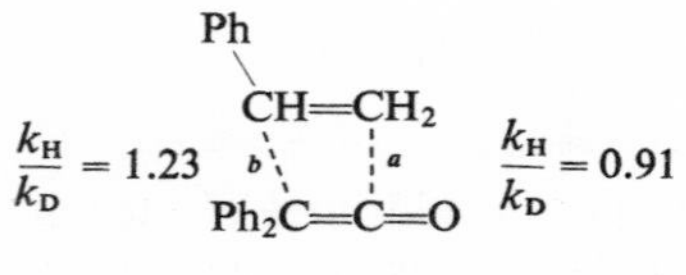

$$\frac{k_H}{k_D} = 1.23 \qquad \frac{k_H}{k_D} = 0.91$$

Fig. 6.8

For the concerted keten additions to fit in with orbital symmetry predictions, they must be $_\pi 2_s + _\pi 2_a$ processes in which the olefin is always found to act as the $_\pi 2_s$ component and the keten as the $_\pi 2_a$ component, and which for some reason are much more favourable than other $_\pi 2_s + _\pi 2_a$ cycloadditions.

Woodward and Hoffmann have suggested that there is a favourable secondary interaction between the highest occupied olefin orbital and the vacant orthogonal π^* antibonding orbital of the keten molecule. This sufficiently stabilises the transition state for the allowed $_\pi2_s + _\pi2_a$ concerted mode of addition (fig. 6.9) so that this process occurs in preference to a stepwise alternative.[1,27]

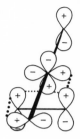

Fig. 6.9

Another way of looking at keten cycloadditions is to consider the reactions as six-electron cycloadditions, either $_\pi2_s + _\pi2_s + _\pi2_s$ or $_\pi2_s + _\pi2_a + _\pi2_a$. This takes account of the participation of the second keten π bond in the reactions (fig. 6.10).

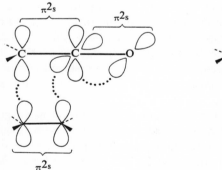 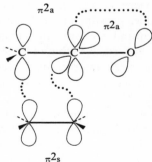

Fig. 6.10

As in other cycloadditions, steric factors can make one mode of $_\pi2_s + _\pi2_a$ addition more favourable than another. Thus the addition of monochloroketen to cyclopentadiene gives the adduct (30) with an *endo*-chlorine.[28]

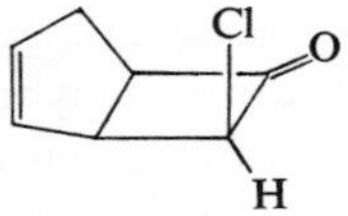

(30)

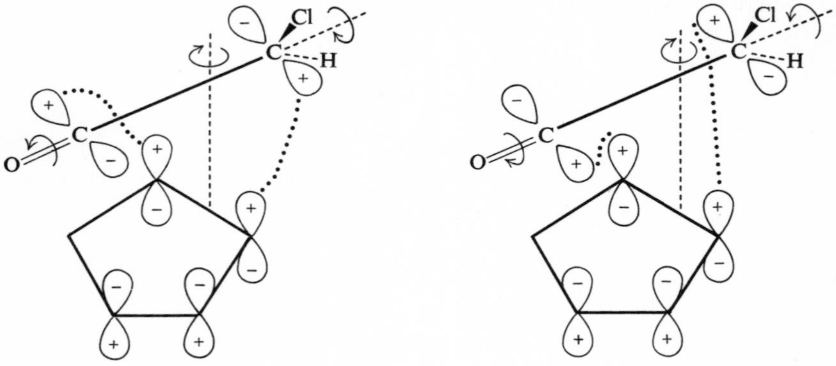

(a)   (b)

Fig. 6.11

It is assumed that the initial approach occurs with the larger of the methylene substituent groups as far from the cyclopentadiene ring as possible. Twisting in the directions shown in fig. 6.11(a) involves better overlap and minimises interaction between the keten and cyclopentadiene hydrogen, whereas that of fig. 6.11(b) involves less efficient overlap and greater steric interaction. In general, the larger keten substituent goes predominantly into the *endo-* position in the adducts with cyclopentadiene.[28]

Huisgen has shown that dimerisation of substituted ketens to give 1,3-cyclobutanediones is also a concerted process where some asymmetry in the transition state leads to slight charge build-up; this explains the observed head-to-tail orientation (fig. 6.12).[29] Keten itself dimerises differently (6.18) to give the lactone (31).

Since keten dimerisation and cycloaddition to olefins can be concerted, it is tempting to speculate that all cumulenic systems may undergo concerted cycloaddition since they have an orthogonal vacant π^* antibonding orbital. Such cycloadditions are not all concerted, however. The ability of the π^* orbital to interact depends on its energy relative to

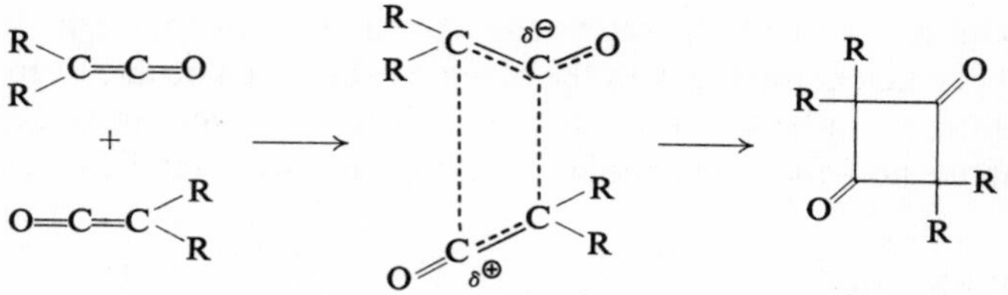

Fig. 6.12

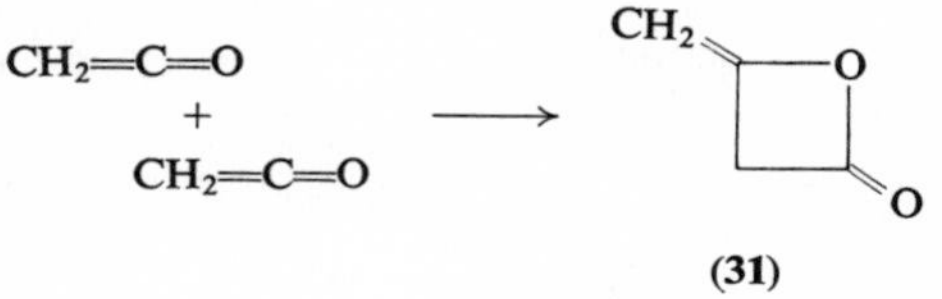

(6.18)

(31)

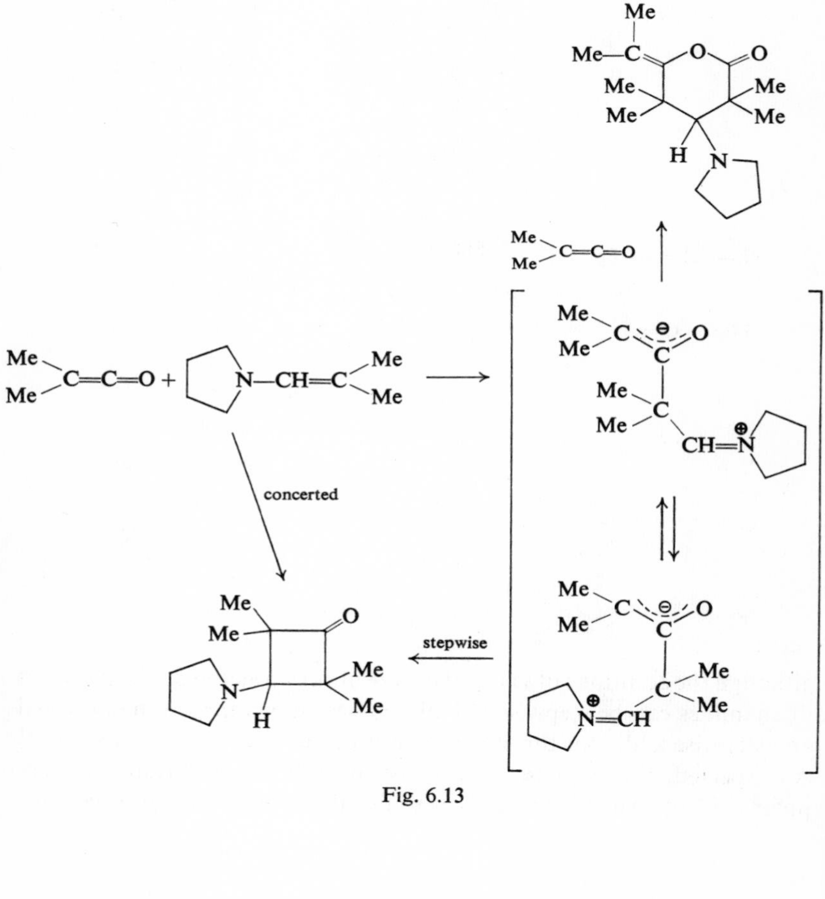

Fig. 6.13

the π orbital of the olefin. In ketens the π^* orbital is exceptionally low-lying so that efficient mixing with the olefin π orbital can occur. Even so, keten additions can be stepwise where the zwitterionic intermediates in the alternative non-concerted mode are particularly well stabilised. Thus,

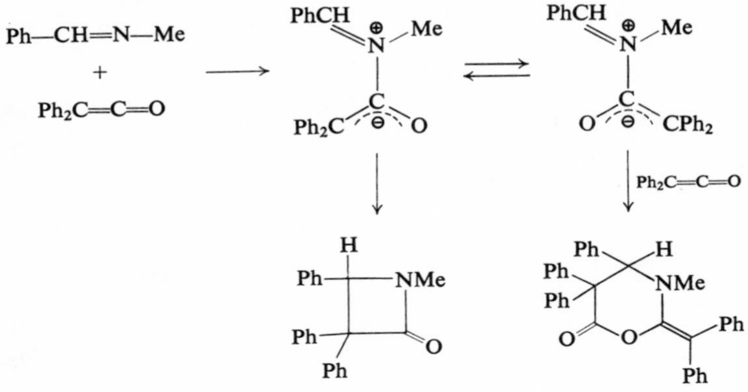

Fig. 6.14

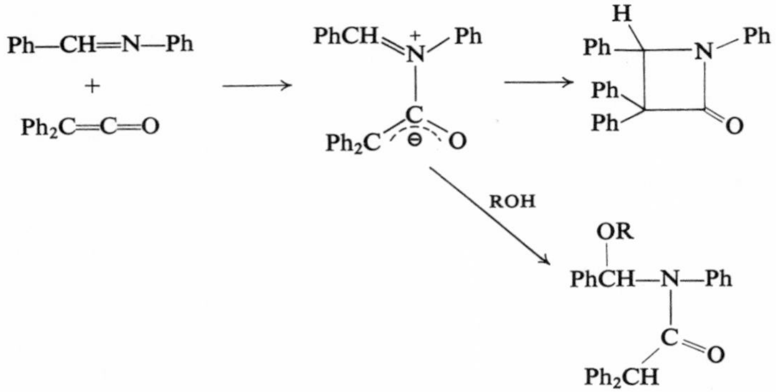

Fig. 6.15

although the additions of vinyl ethers are always concerted, the additions of enamines can be stepwise.[30] Huisgen has shown that both concerted and stepwise additions through a zwitterion can occur simultaneously.[31] As expected, the stepwise process becomes more important in more polar solvents which can better stabilise the zwitterionic intermediate.

It is estimated that the addition of dimethylketen to N-isobutenyl-pyrrolidine is 57 per cent stepwise in acetonitrile but only 8 per cent in cyclohexane. That enamines undergo stepwise addition whereas addition to vinyl ethers is concerted is reasonable since the less electronegative nitrogen atom is better able to accommodate the positive charge in a zwitterionic intermediate than is oxygen. Evidence for the stepwise addition comes from the isolation of 2:1 six-membered ring adducts, as shown in fig. 6.13.[31]

The addition of ketens to Schiff bases is also stepwise; 2:1 adducts have been isolated[32a] (fig. 6.14). The intermediate zwitterions in such reactions have also been trapped by water and methanol[32b] (fig. 6.15) and by sulphur dioxide.[32c]

Very little systematic investigation into the mechanism of the cyclo-additions of other heterocumulenes has yet been undertaken. It is possible that other heterocumulenes can undergo cycloadditions by the allowed concerted $_\pi 2_s + _\pi 2_a$ mode. On the other hand, since all hetero-cumulenes have an electrophilic central atom and nucleophilic terminal atoms, stepwise mechanisms through 1,4-dipolar intermediates are also possible.[4a] Frequently the ionic intermediates are highly stabilised and in many cases the ionic mechanism almost certainly applies. Evidence for this is the trapping of the 1,4-dipolar intermediates, as, for example, in the reaction of ketens with carbodi-imides to give azetidinones (32). If water is added the intermediate dipole (33) is trapped as (34). Control experiments show that the azetidinones are not converted to (34) by reaction with water.[33]

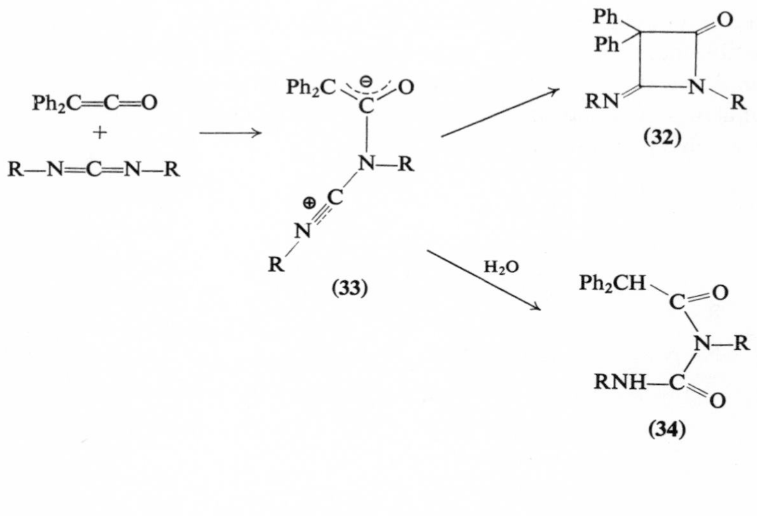

In summary, concerted $_\pi 2_s + _\pi 2_a$ additions are feasible for cumulenes where the π^* orbital is of low energy and where the reactants are such that formation of intermediates is not too favourable.

Other examples of heterocumulene cycloaddition. Examples of heterocumulenes acting as dienophiles are shown in equations 6.19 and 6.20. Because they are such unsymmetrical dienophiles, it is quite likely that many of these $4 + 2$ cycloadditions occur stepwise. Similarly they can form part of a diene component (6.21).

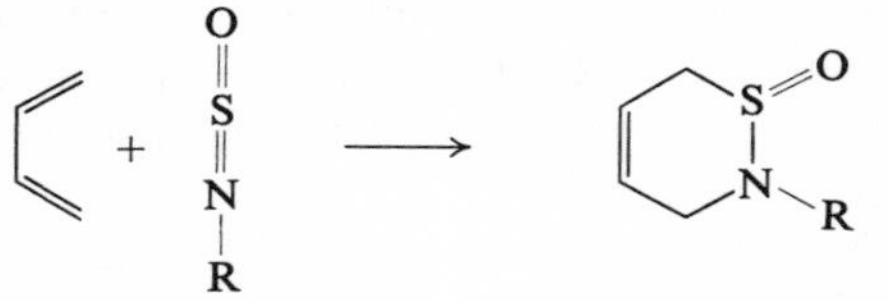

$$\text{(6.19)}$$

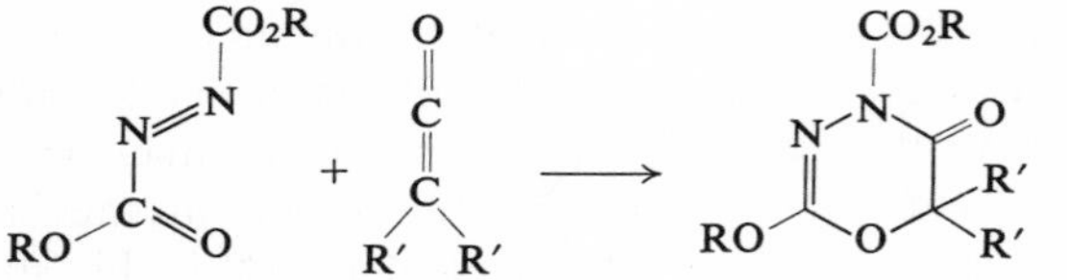

$$\text{(6.20)}$$

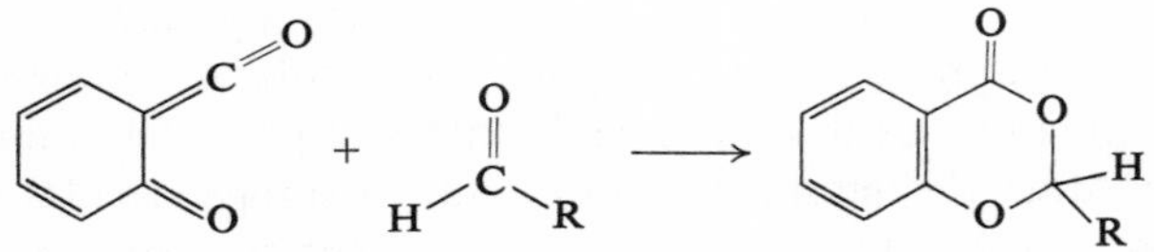

$$\text{(6.21)}$$

Heterocumulenes are usually efficient 1,3-dipolarophiles. Whether these cycloadditions are concerted is very much open to question since ionic intermediates would often be highly stabilised. Because of the possible combinations of heteroatoms in 1,3-dipoles and in heterocumulenes the synthetic potential of 1,3-dipolar additions to heterocumulenes is immense. Examples of some available ring systems are shown in equations 6.22–6.25.

$$\overset{\ominus}{C}H_2-N\overset{\oplus}{=}N + R_2C=C=O \longrightarrow \qquad \text{(6.22)}$$

$$R_2\overset{\ominus}{C}-N-\overset{\oplus}{N}-R + R-N=C=O \longrightarrow \qquad \text{(6.23)}$$

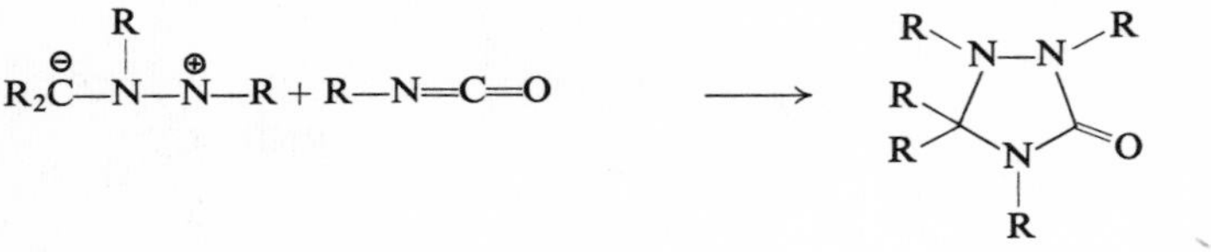

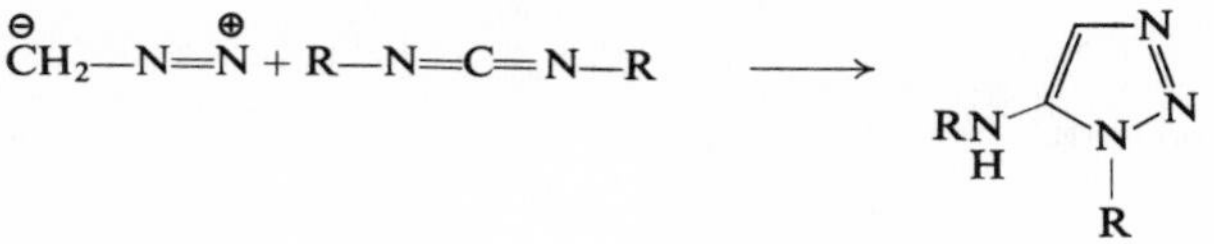

$$\overset{\ominus}{CH_2}-N\overset{\oplus}{=}N + R-N=C=N-R \longrightarrow \qquad (6.24)$$

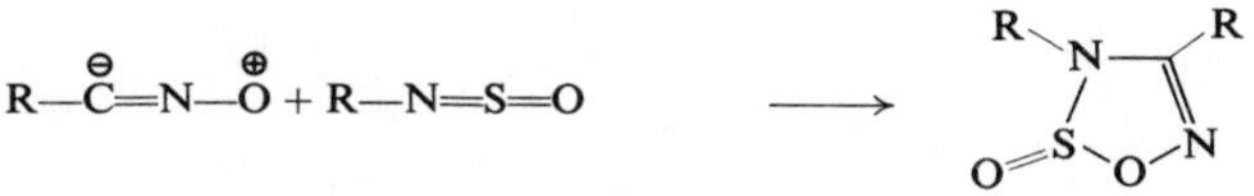

$$R-\overset{\ominus}{C}=N-\overset{\oplus}{O} + R-N=S=O \longrightarrow \qquad (6.25)$$

Heterocumulenes readily undergo $2 + 2$ cycloadditions to form a variety of four-membered heterocycles. The more reactive heterocumulenes tend to dimerise readily and this often leads to difficulty in their isolation (6.26–6.28).

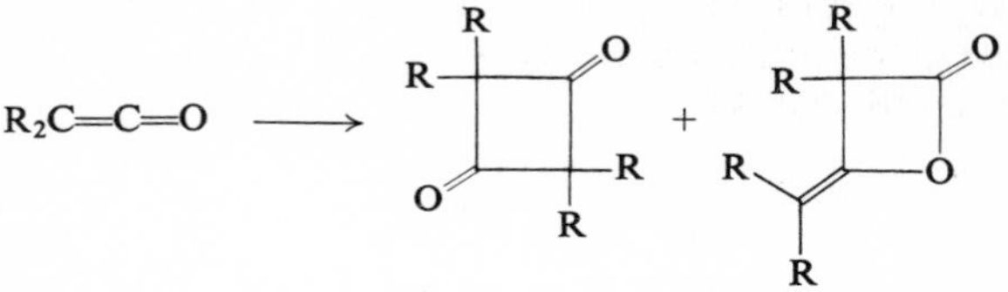

$$R_2C=C=O \longrightarrow \qquad + \qquad (6.26)$$

less commonly

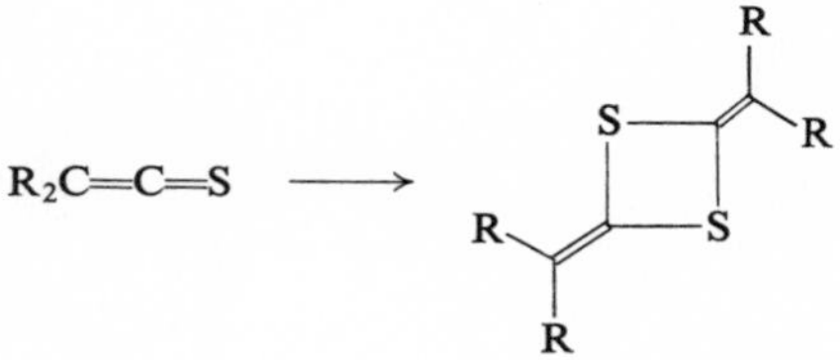

$$R_2C=C=S \longrightarrow \qquad (6.27)$$

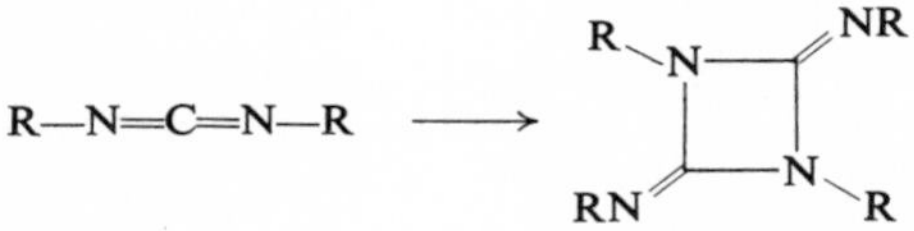

$$R-N=C=N-R \longrightarrow \qquad (6.28)$$

Cycloadditions of one heterocumulene to another also occur readily. In many cases the resulting cycloadducts are unstable and fragment so that overall interchange reactions occur. By selective removal of one component this type of reaction can be made synthetically useful; an example is the formation of unsymmetrical carbodi-imides from iso-cyanates and symmetrical carbodi-imides (6.31).[22,34]

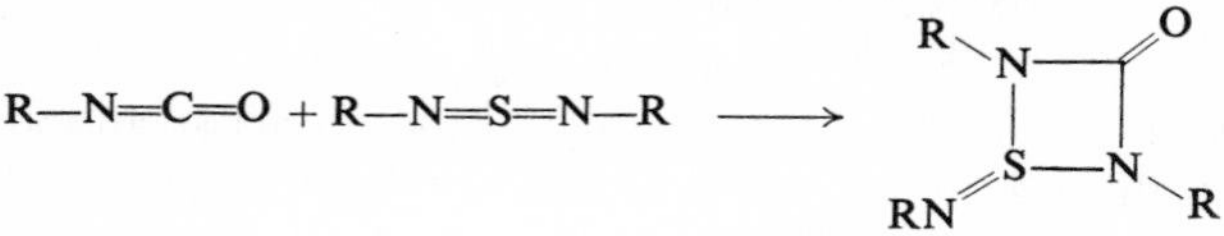

$$R-N=C=O + R-N=S=N-R \longrightarrow \qquad (6.29)$$

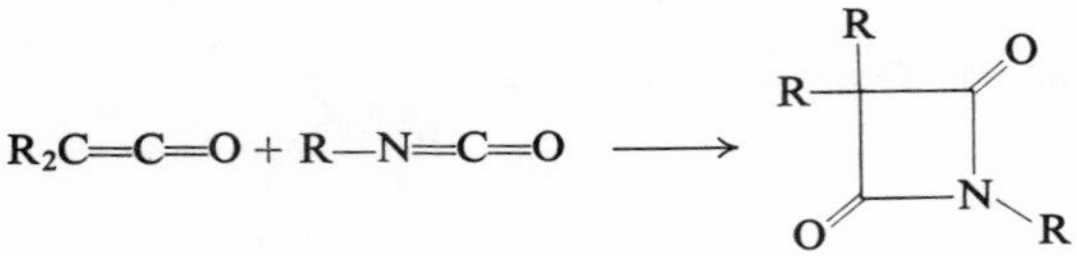

$$R_2C{=}C{=}O + R{-}N{=}C{=}O \longrightarrow \qquad\qquad (6.30)$$

$$R^1{-}N{=}C{=}O + R^2{-}N{=}C{=}N{-}R^2 \rightleftharpoons \qquad\qquad \rightleftharpoons$$

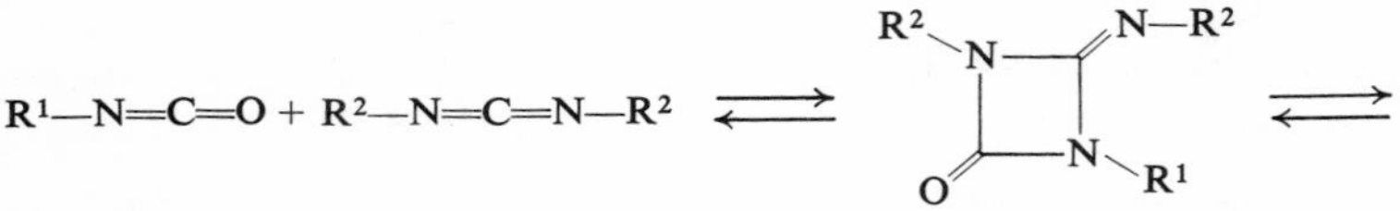

$$R^2{-}N{=}C{=}O + R^1{-}N{=}C{=}N{-}R^2 \quad (6.31)$$

2 + 2 Cycloadditions to a great variety of other multiple bonds are known. Additions of heterocumulenes to C=C, C≡C, C=O, C=N, C=S, N=O, N=N, N=S, S=O, P=C, P=O, P=N and P=S have been reported (6.32–6.35).[22]

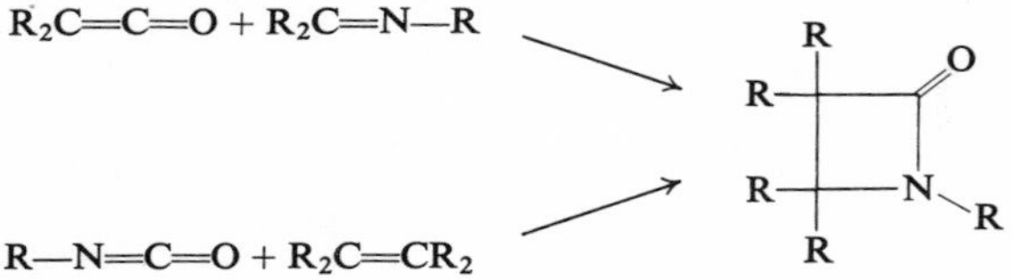

$$R_2C{=}C{=}O + R_2C{=}N{-}R$$

$$R{-}N{=}C{=}O + R_2C{=}CR_2 \qquad\qquad (6.32)$$

$$R_2C{=}C{=}O + R{-}N{=}O \longrightarrow \qquad\qquad (6.33)$$

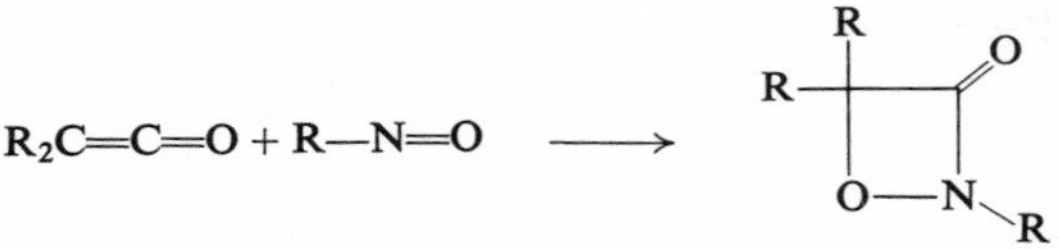

$$R{-}N{=}C{=}O + R_2C{=}N{-}R \longrightarrow \qquad\qquad (6.34)$$

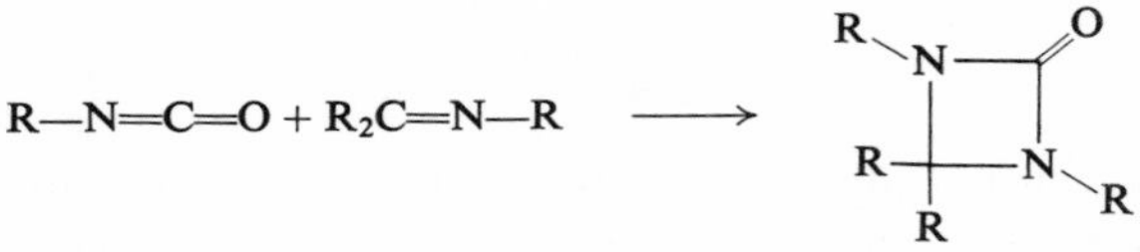

$$R_2C{=}C{=}O + \quad R{-}N{=}N{-}R \longrightarrow \qquad\qquad (6.35)$$

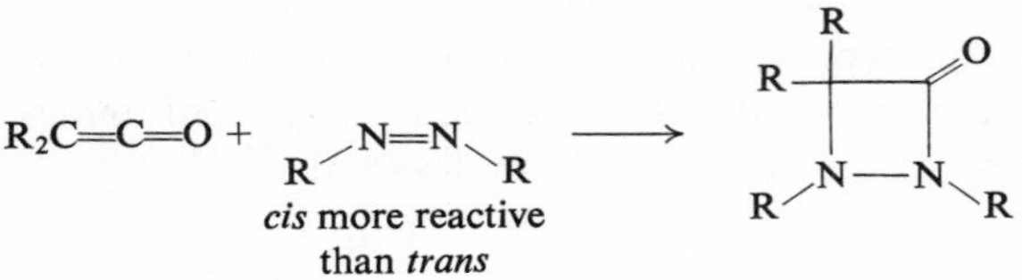

cis more reactive
than *trans*

Cycloadditions of allenes.[4b,13a] On heating, allenes dimerise and also undergo cycloadditions to activated olefins, giving methylenecyclobutanes. In the past these reactions have been assumed to involve di-radical intermediates, a view which fitted in well with initial orbital symmetry predictions that $2 + 2$ cycloadditions must be stepwise. The orientations of the dimerisations and cycloadditions to olefins are consistent with processes which proceed through the most stabilised diradical intermediate. In allene cycloadditions, initial bond formation is generally assumed to occur to the central allene carbon since this leads to an allyl stabilised radical (after 90° rotation of one of the methylene groups). However, because such rotation must precede appreciable stabilisation, it seems unlikely that this effect alone determines the site of initial bond formation.

Head-to-head dimers (35) are always formed (only allene itself forms a minor amount of head-to-tail dimer) and the additions in fig. 6.16 are clearly those resulting from the most stabilised diradical.[35] Other characteristics of the reactions are similar to those of other $2 + 2$ radical cycloadditions (§ 6.3).

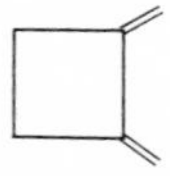

(35)

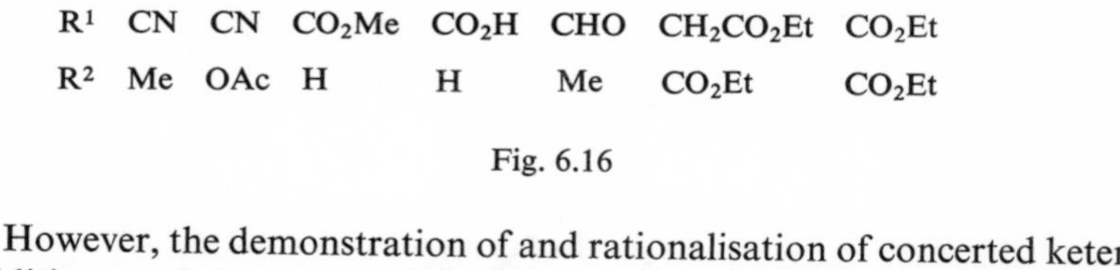

R^1	CN	CN	CO_2Me	CO_2H	CHO	CH_2CO_2Et	CO_2Et
R^2	Me	OAc	H	H	Me	CO_2Et	CO_2Et

Fig. 6.16

However, the demonstration of and rationalisation of concerted keten additions and the observation of high stereospecificity in certain allene cycloadditions has led to speculation that the additions may be concerted $_\pi 2_s + _\pi 2_a$ processes. Diethyl maleate and fumarate add stereo-specifically to 1,1-dimethylallene without any of the rotation expected from a diradical intermediate. The related addition to acrylonitrile is virtually independent of solvent polarity, thus ruling out a highly stereoselective stepwise reaction through a zwitterion.[36] Optically active adducts are formed in the addition of optically active allenes to olefins, indicating stereospecificity in the allene component also.[37]

Optically active and racemic cyclonona-1,2-diene (36) give dimers whose stereochemistries are precisely those expected for the allowed $_\pi2_s + {}_\pi2_a$ addition.[38]

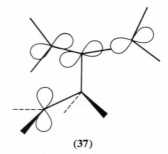

Although these results appear to be consistent with a concerted $_\pi2_s + {}_\pi2_a$ mechanism for allene cycloadditions, some workers maintain that the reactions are stepwise, the stereospecificity of the second bond formation being attributed to other factors. For example an intermediate such as (37) may be so sterically crowded that bond rotations are unable to compete with ring closure.

(37)

A concerted addition is likely to be less favourable than it is with ketens since the π^* orbital of allenes is energetically less accessible for efficient secondary interaction.

The main evidence for an intermediate comes from isotope studies.[39] The dimerisation of tetradeuterioallene occurs at the same rate as the dimerisation of undeuterated allene, suggesting that only bond formation between the central carbon atoms is important in the transition state. Very small or negligible intermolecular secondary deuterium isotope effects are also observed in the addition of 1,1-dideuterioallene to 1,1-dichloro-2,2-difluoroethylene and to acrylonitrile. On the other hand, the dimerisation of 1,1-dideuterioallene shows an intramolecular kinetic isotope effect $k_H/k_D = 1.14$. Deuterium is found predominantly in the methylene groups rather than in the cyclobutane ring indicating that ring closure to CH_2 is faster than to CD_2. Thus the reaction must involve an intermediate which differentiates between two modes of ring closure in a step which is not overall rate determining.

The mechanism of allene cycloaddition is therefore not finally settled.

Cycloadditions of vinyl cations. Vinyl cations have the structure shown in fig. 6.17. They are formed by addition of electrophiles to acetylenes or allenes. The vacant p-orbital orthogonal to the olefin π bond is ideally suited for stabilisation of the transition state for concerted $_\pi 2_s + _\pi 2_a$ addition by overlap with the HOMO of the other component (fig. 6.18). They can therefore be considered as the prototypes for concerted cumulene additions; the vacant p-orbital plays the same role as the π^* orbital of the cumulene.[1] In addition to the stabilisation discussed above, since vinyl cations have an sp-hybridised carbon,

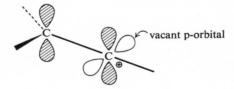

Fig. 6.17

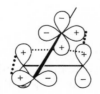

Fig. 6.18

these additions are less susceptible to the unfavourable steric interactions which inhibit the $_\pi 2_s + {}_\pi 2_a$ addition of one olefin to another.

Reactions which can be rationalised by the facile $2 + 2$ addition of vinyl cation intermediates[40] are the formation of the cyclobutane (38) from allene and HCl $(6.36a)$,[41] and of the cyclobutene (39) from but-2-yne and chlorine $(6.36b)$.[42]

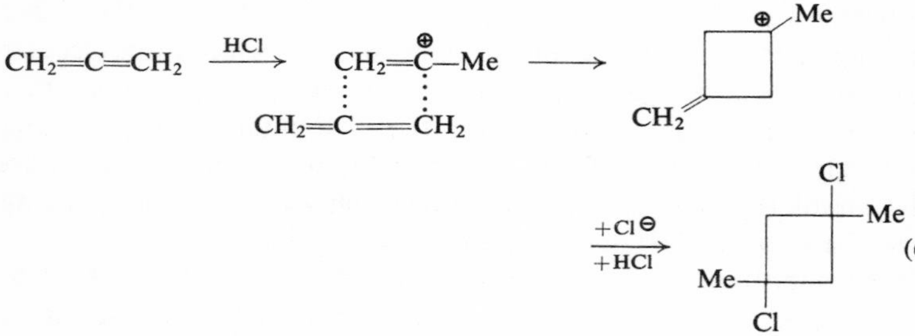

(6.36a)

(38)

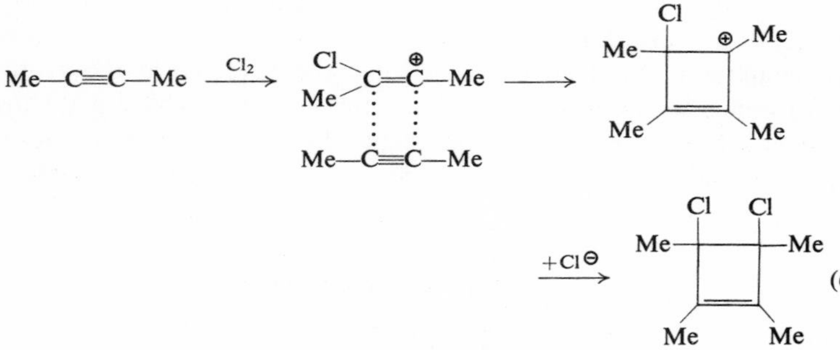

(6.36b)

(39)

6.5. Retro-2 + 2 additions (fig. 6.19)

Fig. 6.19

If concerted, retro-2 + 2 cycloadditions would have to be $_\sigma 2_s + {}_\sigma 2_a$ and therefore involve considerable twisting of the four-membered ring (fig. 6.20). The concerted reaction would be expected to be more energy demanding than a stepwise reaction involving cleavage of one σ bond

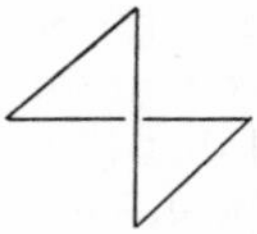

Fig. 6.20

to give a 1,4-diradical. In line with this, cleavage of cyclobutanes requires high temperatures and shows the characteristics of stepwise processes. In some cases, however, there is evidence that the unfavourable $_\sigma 2_s + _\sigma 2_a$ mode does compete.

The reactions have high activation energies (~ 60 kcal mol^{-1}, ~ 250 kJ mol^{-1}).[43] They also show fairly large positive entropies of activation indicative of a large release of ordering in the transition state. This is expected for breaking of one σ bond to give a 'floppy' intermediate but not for simultaneous breaking of two σ bonds, where the potential fragments are still highly ordered with respect to each other in the transition state. For example, concerted retro-Diels–Alder reactions generally show $\Delta S^{\pm} \sim 0$. Cleavage of substituted cyclobutanes proceeds to give mainly fragments derived from the most stable diradical (fig. 6.21) and those cyclobutanes which give more stabilised diradicals cleave more easily. Thus isopropenylcyclobutane (**40**) decomposes with lower activation energy than does isopropylcyclobutane (**41**).[43]

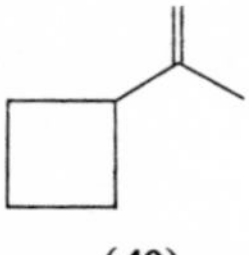

(**40**) 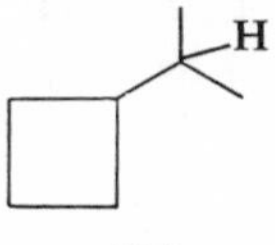   (**41**)

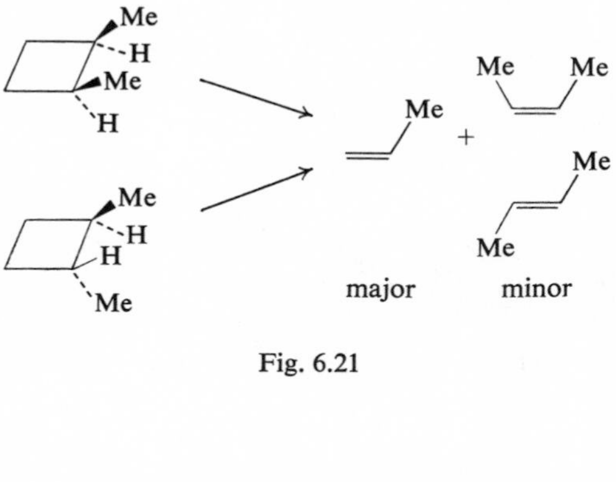

major   minor

Fig. 6.21

The small proportion of butenes formed from *cis*- and *trans*-dimethyl-cyclobutanes (fig. 6.21) shows that some loss of configuration has occurred.[44] It could be argued that since a concerted $_\sigma2_s + _\sigma2_a$ process involves inversion of one fragment, such stereochemical results are also consistent with concerted cleavage in which inversion either of the butene or of the ethylene fragment occurs (fig. 6.22).

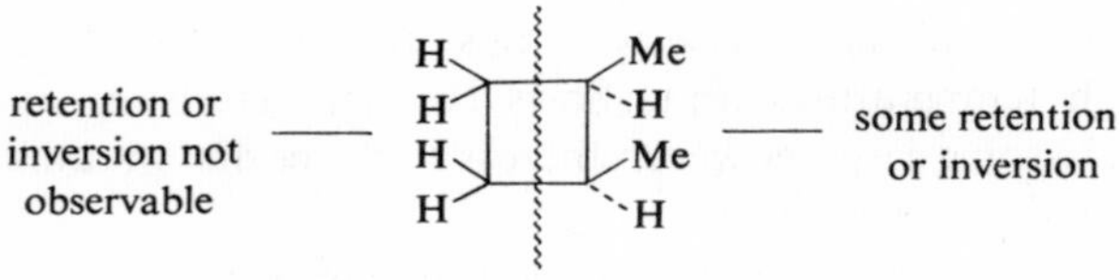

retention or   some retention
inversion not   or inversion
observable

Fig. 6.22

However, systems have been chosen in which the configuration of one of the olefins produced is fixed as *cis* by virtue of it being in a small ring. The stereochemistry of the other fragment is then studied. For a concerted $_\sigma2_s + _\sigma2_a$ reaction its configuration should be completely inverted. The *trans*-dimethylbicycloheptane (**42**) gives but-2-ene which is 75 per cent *trans* (6.37). The *cis* isomer (**43**) gives equal amounts of *cis*- and *trans*-butene. This is just as expected for a diradical intermediate in which cleavage of the second σ bond competes with bond rotation; the driving force for rotation is greater in the diradical in which two methyl groups are *cis*.[45]

The bicyclo-octane (**44**) however gives cyclohexene and dideuterio-ethylene which is 57–62 per cent *trans* (6.38). A diradical intermediate

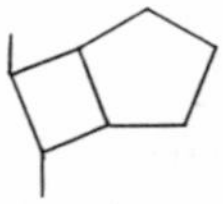

(**42**) $\xrightarrow[\text{gas phase}]{410\text{–}450°C}$ + but-2-ene   (6.37)

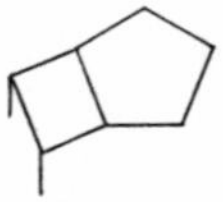

(**43**)

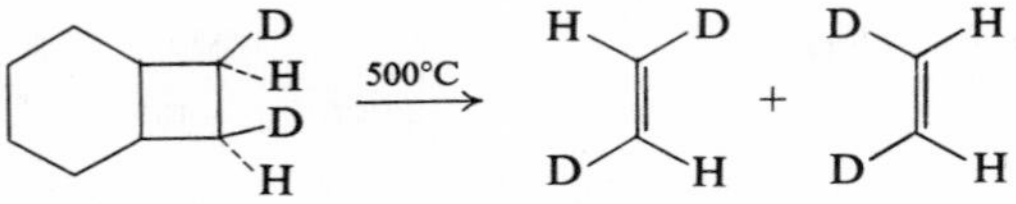

(**44**) $\xrightarrow{500°C}$ +   (6.38)

57–62%

in which bond rotation reached equilibrium before cleavage of the second bond, should give 50 per cent *cis* and 50 per cent *trans*; if cleavage occurred before complete equilibration then *cis* should predominate.[46] These results suggest, therefore, that the concerted $_\sigma 2_s + _\sigma 2_a$ process is competing to some extent with the stepwise diradical route. (Note that in this case the diradical would be less stabilised.)

Retro-2 + 2 cycloadditions of bicyclohexanes **(45)–(47)** to give diallyls leads to a product distribution which appears to be largely consistent with a $_\sigma 2_s + _\sigma 2_a$ process (6.39 and 6.40).[47] However, the facts

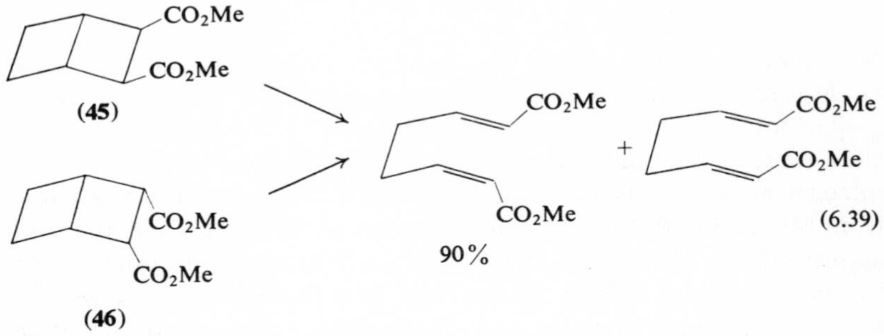

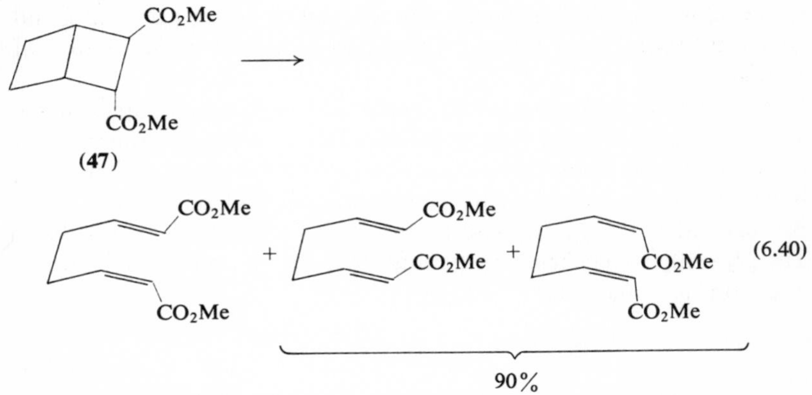

that some product of the wrong stereochemistry is formed and that **(45)** and **(46)** give an identical product distribution, are taken as evidence for a stepwise diradical process with initial cleavage of the strained central bond. Concerted $_\sigma 2_s + _\sigma 2_a$ modes for decomposition of **(45)** and **(46)** involve very different steric interactions and are therefore most unlikely to give the same product distribution. This is therefore an example of a stepwise reaction which largely gives the products expected from a concerted reaction, but only because other factors control the stereochemistry of the stepwise process.[47]

The disallowed nature of retro-2 + 2 addition is underlined by the remarkable stability of the highly strained diazetidine (**48**). Fragmentation of (**48**) to norbornene and nitrogen is estimated to release

(48)   **(49)**

60 kcal mol^{-1} (250 kJ mol^{-1}). In spite of this, decomposition only occurs on heating and the activation parameters are $\Delta H^+ \simeq 34$ kcal mol^{-1} (140 kJ mol^{-1}), $\Delta S^+ \simeq 3$ kcal mol^{-1} (12 J mol^{-1} K^{-1}). This is in striking contrast to compound (**49**) which is far less strained but which loses nitrogen as fast as it is formed, even at $-78°$C. The rates of nitrogen loss from (**48**) and (**49**) differ by a factor of about 10^{22} and a difference of activation energy of about 18 kcal mol^{-1} (75 kJ mol^{-1}) is estimated for the two processes.[48] This reflects very well the difference between a symmetry-allowed concerted retro-Diels–Alder reaction and a reaction for which either the geometrically favoured mode is disallowed ($_\sigma 2_s + _\sigma 2_s$), or the allowed mode is geometrically very unfavourable ($_\sigma 2_s + _\sigma 2_a$), and which therefore most likely proceeds by diradical stepwise loss of nitrogen.

Bicyclobutanes are formed by irradiation of dienes. In spite of the strain in such molecules they are remarkably resistant to isomerisation to the much more stable butadienes. This high temperature isomerisation, however, appears to be a sterically unfavourable but concerted $_\sigma 2_s + _\sigma 2_a$ process rather than a stepwise reaction with a diradical intermediate. The $_\sigma 2_s + _\sigma 2_a$ stereochemistry is illustrated by the examples shown in equations 6.41 and 6.42.[49]

(6.41)

(6.42)

Since concerted $_\pi 2_s + _\pi 2_a$ addition of certain cumulenes to olefins occurs readily, it might be expected that those retro-2 + 2 additions which lead to a cumulene would be more likely to be concerted. This seems to be borne out by the often ready and stereospecific loss of carbon dioxide from β-lactones (6.43).[50]

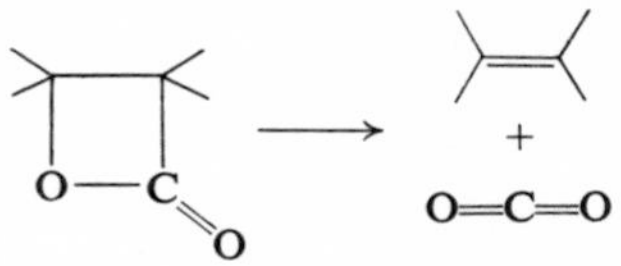

$$(6.43)$$

β-Lactams also lose $HN{=}C{=}O$ stereospecifically on strong heating (6.44) whereas in both the *cis-* and *trans-O*-alkylated lactams (50), under similar conditions, only one bond is cleaved to give the same three products (6.45).[51]

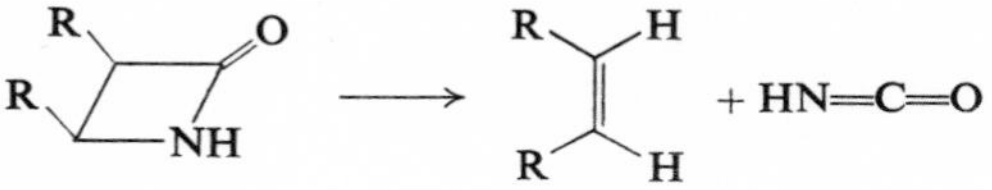

$$(6.44)$$

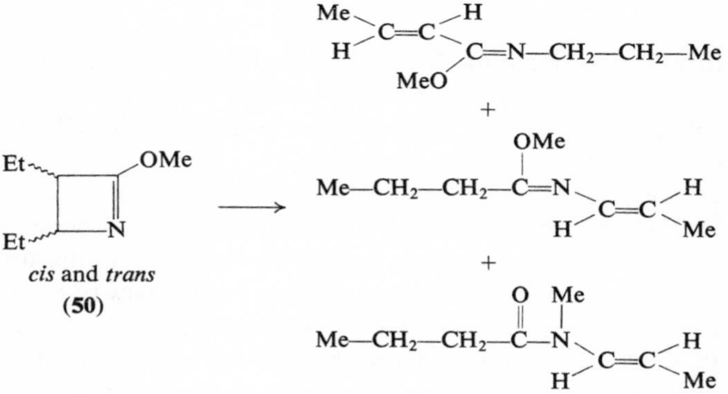

$$(6.45)$$

cis and *trans*
(50)

A particularly interesting retro-2 + 2 addition is the cleavage of 1,2-dioxetanes. If this is a concerted retro-$_\pi 2_s + _\pi 2_s$ reaction, then one of the keto-groups must be formed in an excited state. A case where this occurs is the thermal decomposition of the dioxetane (51) which initiates 'photochemical reactions' of other substrates present (6.46). There are several such examples of photochemistry without light.[52]

The electronically excited species which are responsible for certain types of chemiluminescence and bioluminescence also probably arise by this type of dioxetane cleavage.[53]

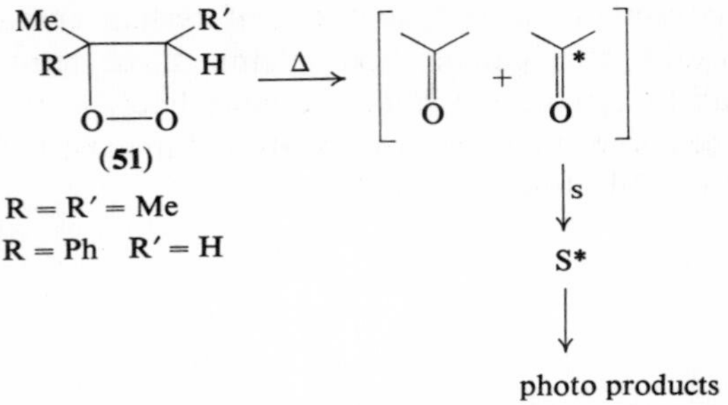

$$
\begin{array}{c}
\text{Me} \diagdown \quad \diagup \text{R}' \\
\text{R} \diagup \quad \diagdown \text{H} \\
\text{O}\!-\!\text{O} \\
\mathbf{(51)}
\end{array}
\;\xrightarrow{\;\Delta\;}\;
\left[\;
\begin{array}{c}
\mathsf{Y} \\
\| \\
\mathsf{O}
\end{array}
\;+\;
\begin{array}{c}
\mathsf{Y}^{*} \\
\| \\
\mathsf{O}
\end{array}
\;\right]
\tag{6.46}
$$

R = R′ = Me
R = Ph   R′ = H

$$\Big\downarrow \text{s}$$

$$\text{S}^{*}$$

$$\Big\downarrow$$

photo products

6.6. Cheletropic reactions.   The cycloaddition of an atom or group X to an olefin to form a three-membered ring and the reverse process constitute a further type of four-electron cycloaddition or elimination. If concerted, such reactions are examples of cheletropic processes (fig. 6.23).

$$
\underset{\bullet\bullet}{\overset{\text{X}}{=\!\!=}}
\;\;\rightleftharpoons\;\;
\underset{\triangle}{\overset{\text{X}}{}}
$$

Fig. 6.23

The forward reaction is limited to the additions of high energy species such as carbenes and nitrenes.[4c] The reverse reaction occurs more widely and a variety of extrusions is known where X is usually a small stable molecule.[54]

The theory of cheletropic reactions was discussed in chapter 4. It was shown that when X has a vacant and a filled orbital the concerted addition or extrusion must proceed by a non-linear cheletropic route. Some of the better known examples of this type of reaction will now be discussed.

Addition of carbenes and nitrenes.[55,4c] Carbenes (:CR$_2$) and nitrenes (:NR) are short-lived reactive intermediates which are electron deficient; they contain a carbon or nitrogen atom with two non-bonding orbitals between which are distributed two electrons. Both the electrons can be in the same orbital, **(52)**, or one electron may be in each, **(53)** and **(54)** (fig. 6.24).

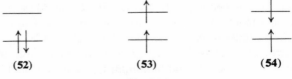

Fig. 6 24

Possible non-linear (55) and linear (56) structures for a carbene :CR$_2$ are shown. The bent structure, which accentuates the difference in energy of the non-bonding orbitals, is likely to be favoured for the singlet ground configuration (52), whereas the excited configurations (53) and (54) are likely in a linear or nearly linear structure such as (56), where the non-bonding orbitals are close or equivalent in energy.

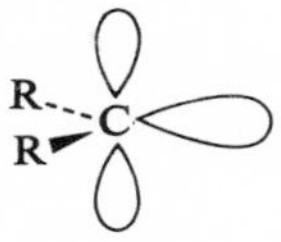

(55)    (56)

Depending on the nature of the substituents, either configuration (52) or configuration (53) can be the actual ground state (the lowest energy state) of the carbene or nitrene. Most ground states are triplets, of type (53), but the singlet configuration (52) is normally the ground state when both substituents have a lone pair (F, Cl, Br, OR, NR$_2$, etc.). This can be envisaged, in valence bond terms, as being due to resonance stabilisation of configuration (52) by the lone pair on the hetero-atom (fig. 6.25).

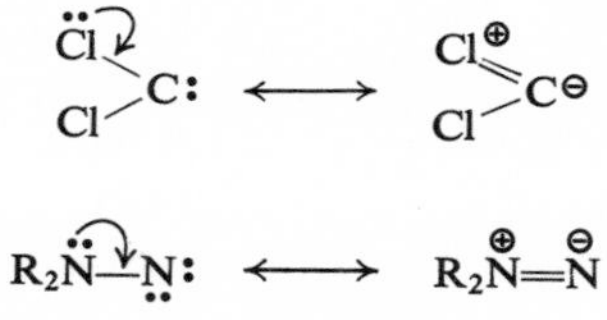

Fig. 6.25

Carbenes and nitrenes add to olefins, to give cyclopropanes and aziridines, respectively. Such reactions are synthetically very useful. These additions are not given by all carbenes and nitrenes: many preferentially undergo rearrangement, fragmentation, insertion, or hydrogen abstraction. Sometimes these reactions can be avoided by using metal catalysis in the decomposition of the carbene precursors, or by using organometallic sources of the intermediates. For example, thermal uncatalysed decomposition of diazoketones gives mainly products of the Wolff rearrangement, but in the presence of a copper catalyst, cycloaddition to an olefin can compete (6.47). In such cases, the cycloadditions probably do not involve free carbenes, but 'carbenoids' – a term loosely used to describe complexed carbenes or carbene-like

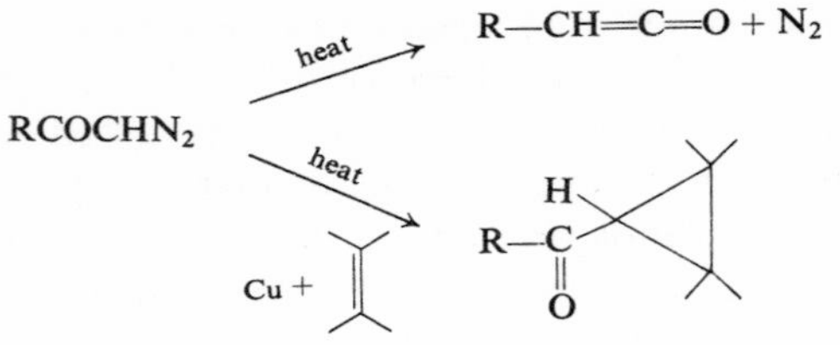

$$R\text{—}CH\text{=}C\text{=}O + N_2$$

$$RCOCHN_2 \qquad (6.47)$$

intermediates. Intermolecular cycloadditions by free, uncomplexed carbenes and nitrenes are limited to relatively few types; the major ones are shown in table 6.3.

TABLE 6.3. *Carbenes and nitrenes which commonly add to olefins*

Carbenes	
methylene	$:CH_2$
dihalogenocarbenes	$:CX_2$ (X = F,Cl,Br)
alkoxycarbonylcarbenes	$:CHCO_2R$
vinylidene carbenes	$:C\text{=}CR_2$
diarylcarbenes	$:CAr_2$
arylhalogenocarbenes	$:CArX$
atomic carbon	$:C:$
Nitrenes	
alkoxycarbonylnitrenes	$:\dot{N}CO_2R$
cyanonitrene	$:\dot{N}CN$
aminonitrenes	$:\dot{N}NR_2$

In some cases, the addition is stereospecific, and in others it is not. Stereospecific addition to olefins is observed with carbenoids, and for carbenes which have been shown independently to have singlet ground states of type (52); examples are dihalogenocarbenes, such as :CCl$_2$ and :CF$_2$, and aminonitrenes (diazenes), :ṄNR$_2$. With species for which the ground state is probably a triplet, (53), the pattern is more complex. These intermediates may be generated in configuration (52), and if they react as such before decay to the ground state, their addition is stereospecific. If they are deactivated by collision before addition, the reaction is non-stereospecific.

This pattern is in accord with the orbital symmetry predictions. Only a species of configuration (52), with a filled and a vacant orbital, can participate in the concerted, non-linear cheletropic mode of addition described in chapter 4. The cycloaddition of species (53) or (54) must be stepwise and is therefore likely to be non-stereospecific. In the particular case of the addition of methylene, :CH$_2$, to ethylene, Hoffmann has carried out molecular orbital calculations which show that the excited configurations (53) and (54) of methylene react with ethylene to produce an excited configuration of cyclopropane (6.48), which is an open three-carbon intermediate (trimethylene) with no barriers to rotation.[56]

$$H_2\dot{C}\cdot \ + \ \begin{matrix} CH_2 \\ \| \\ CH_2 \end{matrix} \longrightarrow \quad H_2\underset{\cdot}{C} \diagup \overset{\overset{\displaystyle H_2}{C}}{} \diagdown \underset{\cdot}{C} H_2 \qquad (6.48)$$

This excited configuration of trimethylene has no particular driving force for ring closure and will therefore presumably undergo rotation about the C—C bonds before collapsing to cyclopropane.

The stereospecificity of the addition therefore depends on whether the species reacts in the electronic configuration (52), with both electrons in one orbital, or in the configuration (53) or (54), with one electron in each. It is *not* primarily the spin state which determines the stereospecificity, but the distribution of the electrons.

An earlier widely accepted explanation for the stereospecificity of the additions of the different types of carbene was based on the principle of the conservation of spin. It was assumed that only spin paired singlet carbenes could add in a concerted manner; triplet carbenes would necessarily produce a triplet diradical intermediate which could only ring close after spin inversion. Since this was assumed to be slow compared to bond rotation, addition of a triplet carbene would be non-stereospecific. This theory, known as the *Skell hypothesis*, has proved to be a very useful practical guide.[57]

Extrusions. The overall picture for extrusion reactions is by no means clear and general conclusions as to the mechanism cannot be made. Extrusions of some fragments, for example sulphur dioxide and nitrous oxide, are stereospecific, whereas others, for example sulphur monoxide,

$$MeCH_2SO_2CHClMe \ \rightleftharpoons \ Me\overset{\ominus}{C}HSO_2CHClMe \ \xrightarrow{\text{slow}}$$

$$\begin{matrix} MeCH—CHMe \\ \diagdown S \diagup \\ O_2 \end{matrix} \ \xrightarrow{\Delta} \ MeCH{=}CHMe + SO_2 \qquad (6.49)$$

$$\text{stereospecific}$$

(57)

are not. It is tempting to speculate that those extrusions which are stereospecific are concerted and are examples of non-linear cheletropic processes, but this may be an oversimplification.

Fragmentation of episulphones and episulphoxides.[58] Episulphones (57) are intermediates in the formation of alkenes from α-chlorosulphones and bases (6.49). They can also be synthesised independently; for example, from a sulphonyl chloride, triethylamine and diazomethane (6.50).

$$RCH_2SO_2Cl + CH_2N_2 + Et_3N \longrightarrow \underset{R}{\overset{SO_2}{\triangle}} + N_2 + Et_3\overset{\oplus}{N}HCl^{\ominus} \qquad (6.50)$$

When they are heated, alone or in solution, episulphones are fragmented stereospecifically to sulphur dioxide and alkenes. This can be rationalised by a concerted, non-linear extrusion of sulphur dioxide. However, it must be emphasised that this is a rationalisation of the observed stereospecificity, not a proven mechanism. Alternative stepwise routes, involving, for example, initial ring expansion of an episulphone to a four-membered ring isomer, have also been proposed.[59] It may well be that different mechanisms operate in different systems. In contrast, the extrusion of sulphur monoxide from episulphoxides is non-stereospecific and therefore most probably occurs stepwise (6.51).[60]

$$\underset{RCH-CHR}{\overset{O}{\overset{\|}{\underset{\diagdown}{S}}}} \longrightarrow RCH{=}CHR + SO \qquad (6.51)$$

$$\text{non stereospecific}$$

Fragmentation of aziridine derivatives. *N*-Nitrosoaziridines are unstable compounds formed from aziridines and nitrosyl chloride at low temperature (6.52). At room temperature they decompose stereospecifically giving nitrous oxide and the olefin.[61] Again the result can be rationalised by the 'non-linear extrusion' mechanism.

$$(6.52)$$

The extrusion of nitrogen from the diazenes (58) is more confusing. The deamination of aziridines (59) and (60), (R = Me) by difluoroamine, which proceeds through the diazene, is stereospecific.[62] However the diazenes (58, R = Ph), generated by oxidation of the *N*-aminoaziridines (61), fragment non-stereospecifically. The *trans* isomer gives 100 per cent *trans*-stilbene; the *cis* gives 85 per cent *trans* and 15 per cent *cis*. It is not yet clear whether the extrusion is non-stereospecific or whether the lack of stereospecificity results from isomerisation of the diazene prior to fragmentation.[63]

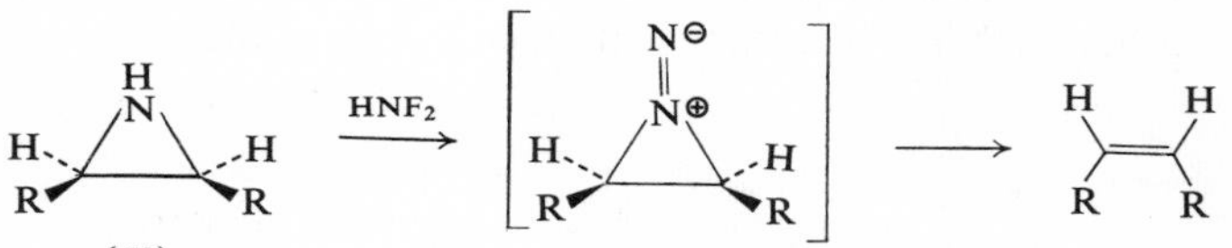

(59)   (58)

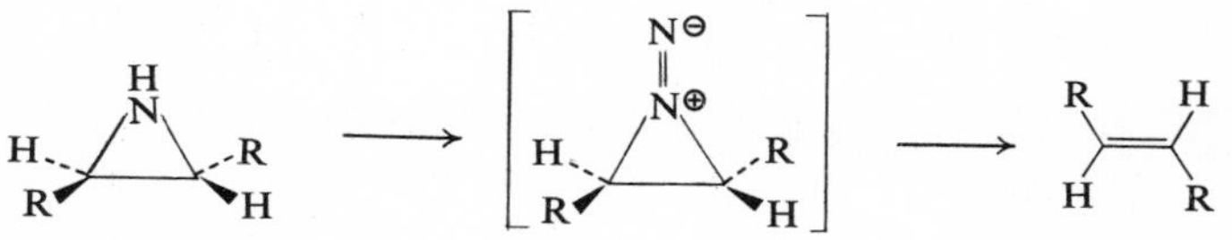

(60)

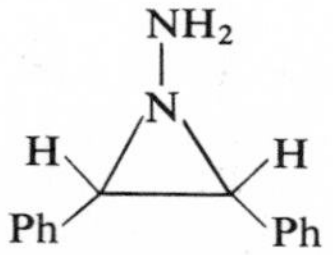

cis and *trans*
(61)

The *N*-aminoaziridines (61) are surprisingly unstable and are fragmented to give an olefin together with a species of stoichiometry H_2N_2, not yet identified.[63]

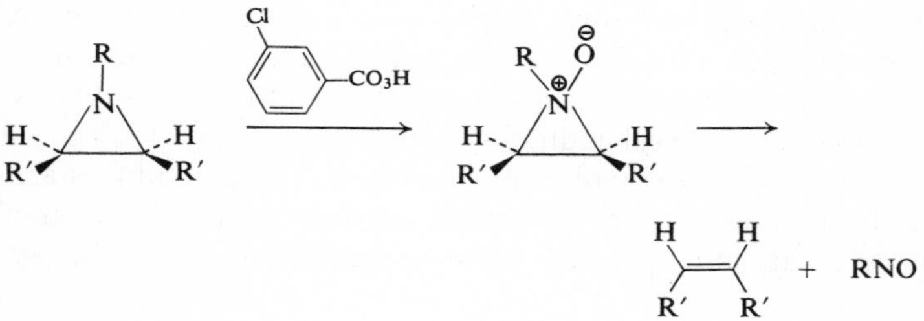

+ RNO   (6.53)

N-Alkylaziridines can be converted stereospecifically to olefins by treatment with *m*-chloroperbenzoic acid. This presumably involves extrusion of a nitroso-compound from an intermediate aziridine *N*-oxide (6.53).[64]

Miscellaneous fragmentations. There are several examples of the fragmentation of cyclopropanes to give carbenes (6.54).[55] This is generally a photochemical reaction. Carbenes have also been produced by the photochemical fragmentation of oxiranes (6.55), and nitrenes have been obtained similarly from oxaziridines (6.56). It is not yet possible to draw general conclusions as to the mechanism of these photofragmentations, although there is evidence that at least some of them are concerted.

$$\text{(cyclopropane)} \xrightarrow{h\nu} \text{:CH}_2 + \text{PhCH}{=}\text{CH}_2 \qquad (6.54)$$

$$\text{(oxirane)} \xrightarrow{h\nu} \text{Ph}\ddot{\text{C}}\text{H} + \text{PhCHO} \qquad (6.55)$$

$$\text{(oxaziridine)} \longrightarrow \text{Ph}\ddot{\text{N}} + \text{Ph}_2\text{CO} \qquad (6.56)$$

6.7. Photochemical 2 + 2 cycloadditions.[65,4b,4c] Photochemical $2 + 2$ cycloadditions have been known for some considerable time but it is only since the 1950s that the synthetic scope of the reaction has been fully exploited.

Although concerted $_\pi 2_s + _\pi 2_s$ addition is photochemically allowed, few photochemically induced $2 + 2$ additions proceed by this mechanism. Triplet excited reactants are definitely involved in many cases reported, where photosensitisers were used. Singlet excited reactants formed by direct irradiation can collapse to lower triplet states in competition with intermolecular collision so that many photochemical cycloadditions are therefore stepwise radical additions.

Examples most likely to involve photoinduced $_\pi 2_s + _\pi 2_s$ addition are the direct irradiation of pure tetramethylethylene[66] and of *cis*- and *trans*-butenes. The latter proceeds stereospecifically as shown in fig. 6.26.[67]

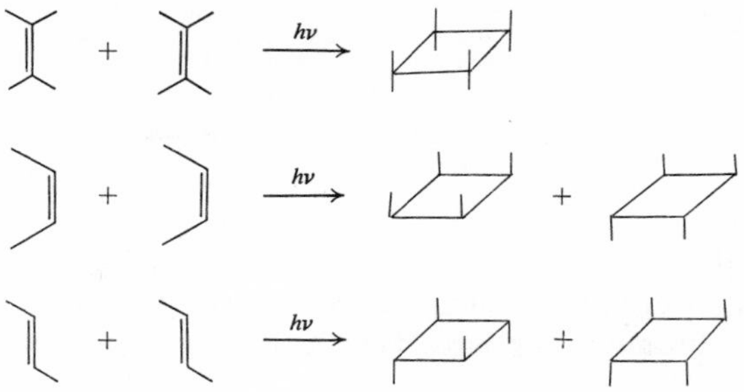

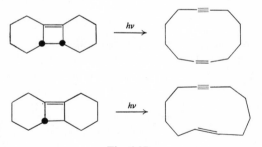

Fig. 6.26

Intramolecular photochemical cycloadditions such as **(62)** → **(63)** may also be concerted since in these cases addition may well compete with relaxation and intersystem crossing. The photochemical retro-

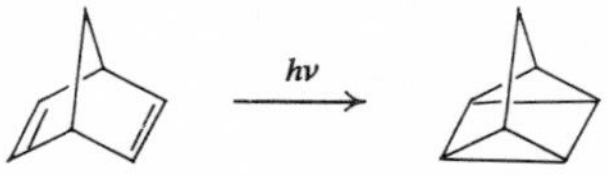

(62)   **(63)**

2 + 2 additions in fig. 6.27 also proceed stereospecifically, suggesting that they are concerted.[68]

Fig. 6.27

There are examples of cycloalkene photodimerisation by direct irradiation but many of these are believed to involve triplet states.[65a] However, the addition shown in equation 6.57 does proceed through a singlet excited state.[69] Other simple olefin dimerisations involve triplet sensitised photoexcitation.

Triplet sensitised photodimerisation of dienes and photoaddition of dienes to monoenes leads generally to 2 + 2 adducts although some 2 + 4

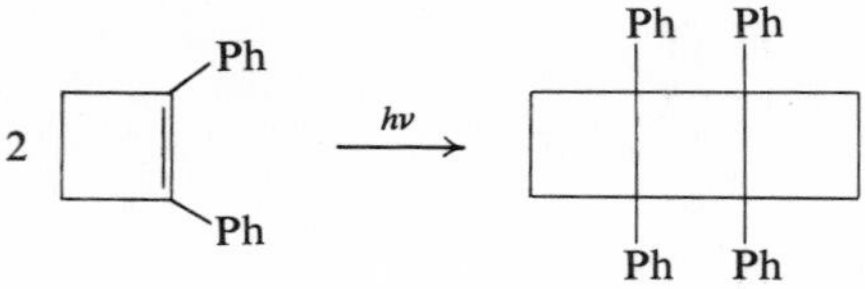

$$2 \quad \boxed{\begin{array}{c} Ph \\ \\ Ph \end{array}} \quad \xrightarrow{h\nu} \quad \boxed{\begin{array}{cc} Ph & Ph \\ \\ Ph & Ph \end{array}} \qquad (6.57)$$

addition is also observed.[65a] Photoadditions of maleic acid derivatives to olefins and to aromatic compounds is a very general route to cyclo-butanes.[65a] The additions have been extended to acetylenes to give cyclobutenes.

On irradiation, cyclopentenones and cyclohexenones dimerise or add to olefins, acetylenes and allenes (6.58 and 6.59).[70,65a] These photo-

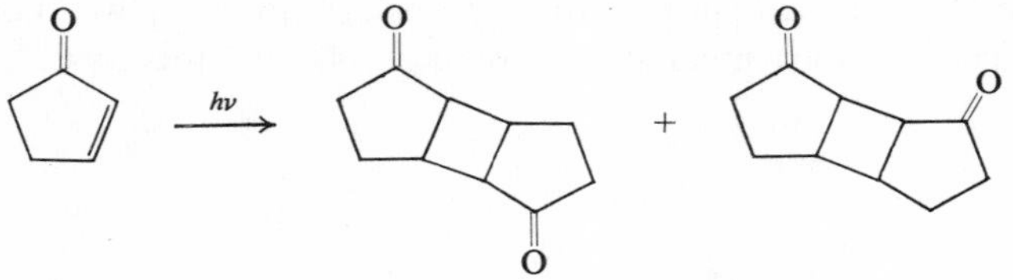

$$\text{(6.58)}$$

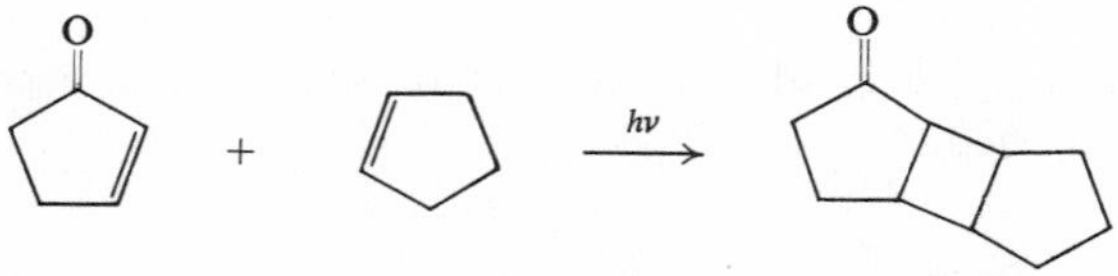

$$\text{(6.59)}$$

additions are believed to occur stepwise and to involve triplet excited enones. They appear to be limited to cyclic enones but otherwise they are quite general and have proved valuable in organic synthesis. A range of

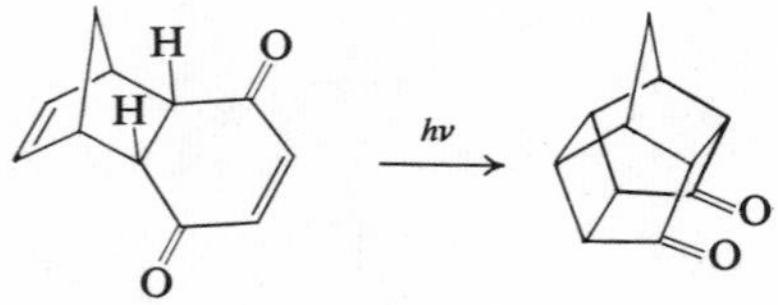

$$\text{(6.60)}$$

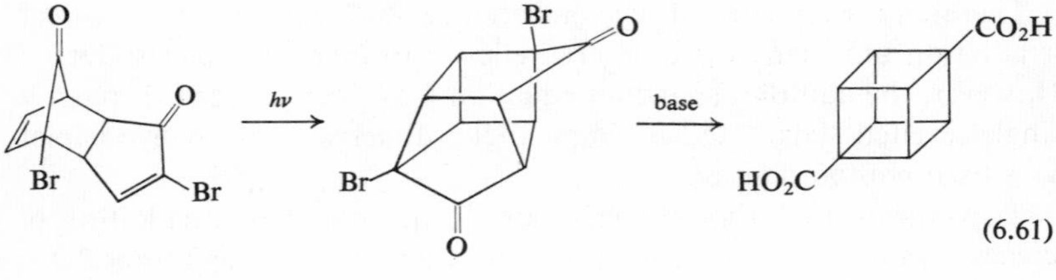

$$\text{(6.61)}$$

novel cage compounds have been produced by intramolecular reactions of this type (6.60, 6.61). Such photoadditions are key steps in Corey's elegant syntheses of caryophyllene (**64**, 6.62) and α-caryophyllene alcohol (**65**, 6.63).[71]

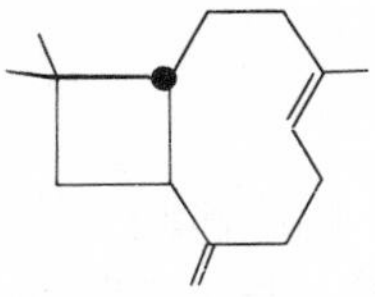

(**64**) 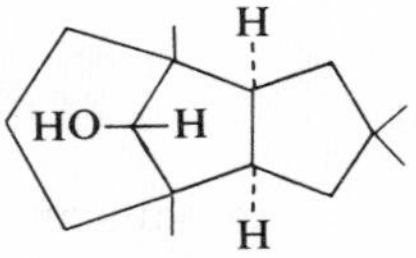   (**65**)

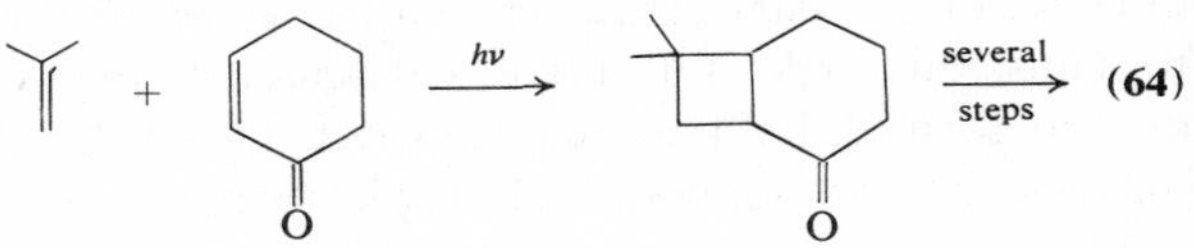

$$\text{(6.62)}$$

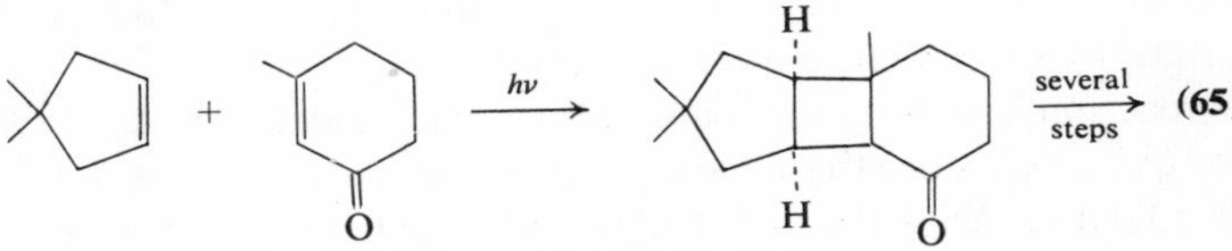

$$\text{(6.63)}$$

Photodimerisation of enones may also have immense biological significance. One of the causes of ultraviolet damage to DNA is believed to be photodimerisation of thymine units in the DNA chains. Thymine itself undergoes a ready photodimerisation (6.64).[72]

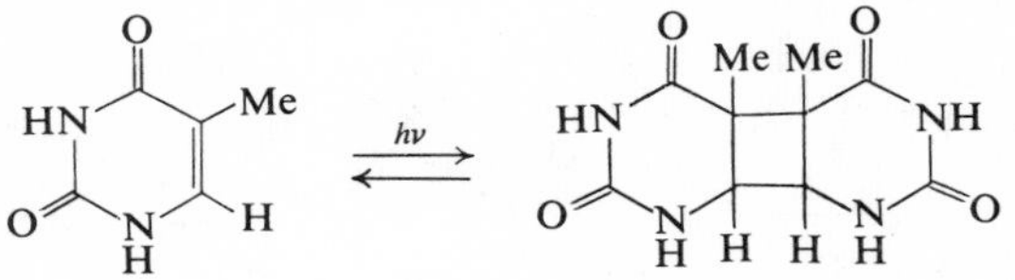

$$\text{(6.64)}$$

Another very general photocycloaddition is that of aldehydes, ketones and quinones to olefins to give oxetanes (6.65, 6.66).[65a,4b,4c] This is

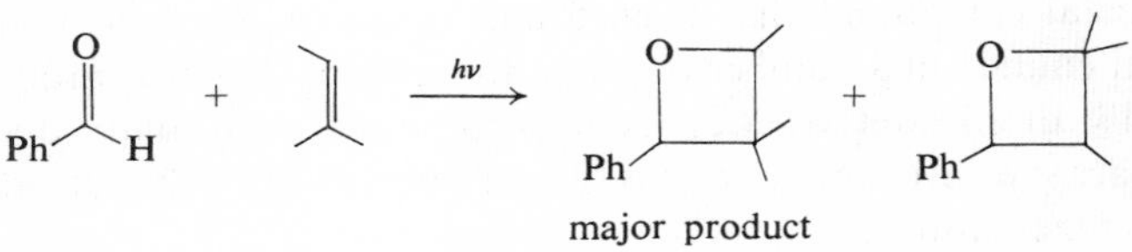

$$\text{(6.65)}$$

major product

(6.66)

known as the Paterno–Buchi reaction, following its discovery by Paterno and the initial investigations into its mechanism by Buchi. The reaction involves excitation of the carbonyl group; the excited singlet rapidly undergoes intersystem crossing to a triplet which then adds to the olefin. Thus, mixtures of isomeric oxetanes, having all possible orientations, are formed, but the oxetane expected from the most stable diradical intermediate predominates. The triplet carbonyl can transfer its energy to the olefin if the latter has a triplet state of suitable energy. Side products from triplet sensitised reactions of the olefins (for example, dimerisation) are therefore often observed.

The reaction is generally efficient for aromatic aldehydes and ketones, and for quinones, since these have lower triplet state energies than most simple alkenes. Aliphatic aldehydes and ketones have higher triplet energies and few examples of such additions are known (those of poly-fluoroalkyl carbonyl compounds are an exception). The carbonyl compounds act as triplet sensitisers for dienes and other conjugated olefins which have low triplet energies. (See, for example, p. 135.)

Photoaddition of carbonyl compounds to acetylenes leads to α,β-unsaturated ketones, presumably via oxetenes (**66**, 6.67).

(6.67)

(66)

6.8. Reactions catalysed by transition metals. The presence of transition metal compounds has been found to divert reactions from their normal courses in a number of cases. In many of these the product of a formally disallowed reaction is formed under very mild conditions. There are several examples of this type of catalysis in 2 + 2 cycloadditions and fragmentations.[73]

A typical example is the dimerisation of norbornadiene which occurs in the presence of iron, nickel and cobalt compounds (fig. 6.28).[73a]

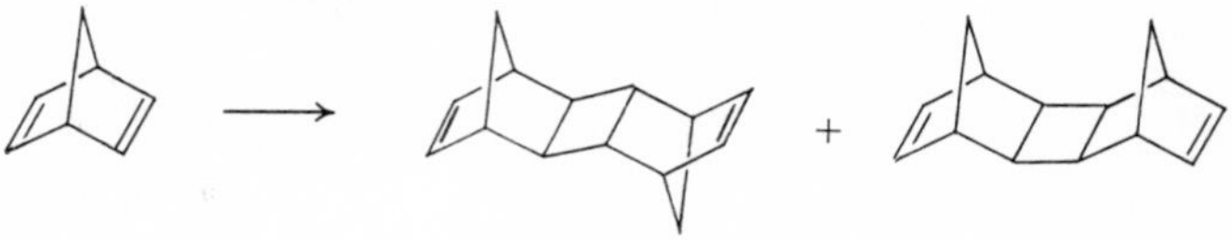

10 : 1

Fig. 6.28

The isomerisation of quadricyclene (**67**) to norbornadiene illustrates a metal catalysed retro-2 + 2 addition (6.68). In this case the thermal reaction has a half life of 14 hours at 140°C, but in the presence of 2 mole per cent of $[Rh^I(norbornadiene)Cl]_2$ the half life is only 45 minutes at −26°C.[74]

(6.68)

(**67**)

An interesting type of reaction which formally combines forward and reverse 2 + 2 additions is the dismutation of olefins (fig. 6.29).[73a] The reaction can be brought about at elevated temperatures in the presence of heterogeneous metal oxide catalysts. It can also take place under very much milder conditions in the presence of homogeneous transition metal catalysts, when it is often called a metathesis.

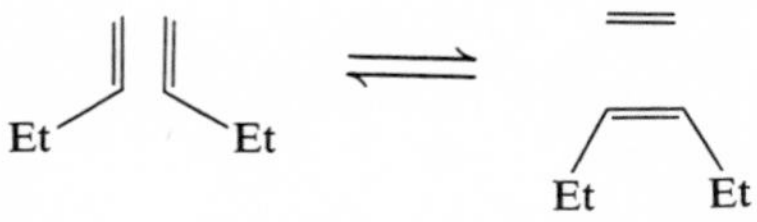

Fig. 6.29

Free cyclobutanes are probably not intermediates in the metathesis; 'quasi cyclobutane' intermediates, in which the components are bonded to the metal, seem more likely.[75] An example of metathesis is the conversion of pent-2-ene into an equilibrium mixture of pent-2-ene, but-2-ene and hex-3-ene in the presence of tungsten catalysts (fig. 6.30). The equilibrium is established within a few seconds at room temperature.

Metathesis also provides a route to catenanes, which have two carbocyclic rings interlocked. Large cyclic olefins (**68**) can combine by this

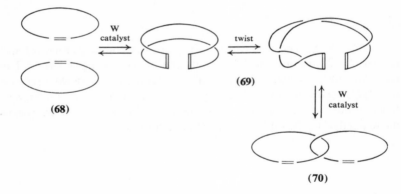

50%   25%   25%

Fig. 6.30

reaction to give conformationally mobile cyclic dienes (**69**). If disproportionation of these dienes, to regenerate the cyclic monoenes, occurs after twisting of the alkyl chains, the two cyclic monoenes may be interlocked (**70**).[76]

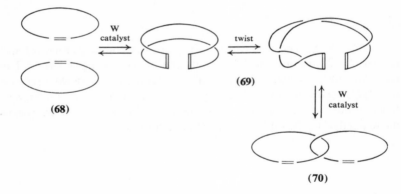

The origin of the catalytic effect of transition metals was discussed in chapter 3. There is increasing evidence that at least some of these catalysed reactions are stepwise and not concerted processes which are allowed because co-ordination to the transition metal changes the orbital symmetry requirements, nor concerted disallowed processes for which the energy barrier is lowered by metal orbitals. Thus, for example, the transformation of cubane to tricyclo-octadiene is catalysed by small amounts of certain rhodium(I) compounds of the type $[Rh^I \text{ diene } Cl]_2$ (6.69).[77]

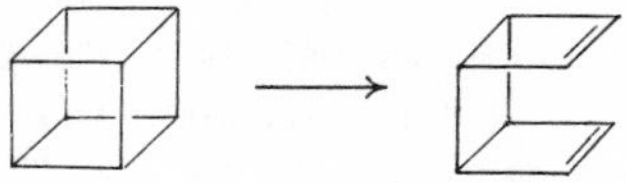

(6.69)

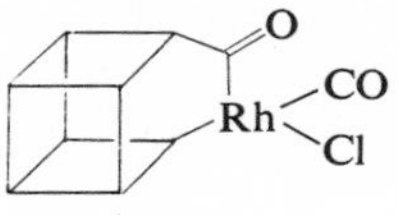

(71)

With stoichiometric amounts of related rhodium complexes $[Rh(CO)_2Cl]_2$, products **(71)** resulting from oxidative addition of cubane to the rhodium complex can be isolated. It therefore seems likely that the catalysed rearrangement of cubane, and related catalysed reactions, proceed in a stepwise manner in which a key step is oxidative addition (6.70).[77]

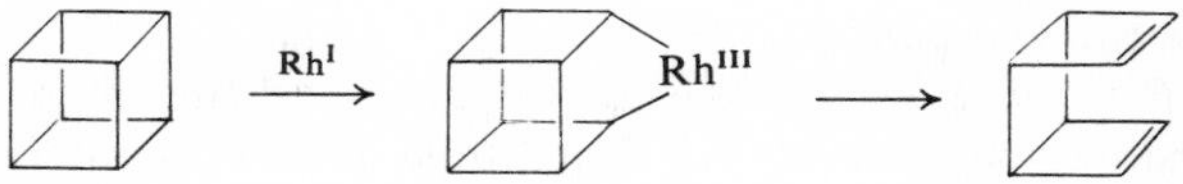

(6.70)

6.9. Other cycloadditions. In this section, cycloadditions involving more than six electrons are discussed. These are far less common than those involving six or fewer electrons and very little detailed mechanistic work of the type done on the Diels–Alder reaction or 1,3-dipolar addition has yet been carried out. The concerted or stepwise nature of these reactions is therefore largely a matter for speculation. The general success of the Woodward–Hoffmann rules in rationalising the occurrence of concerted or stepwise reactions has been so great however, that frequently the mere observation of an allowed process is taken to imply a concerted reaction; and similarly disallowed reactions are assumed to be stepwise. This is too sweeping a generalisation.

Concerted cycloadditions of open chain polyenes leading to large rings are unlikely from entropy considerations. An alternative concerted mode leading to a smaller ring will normally be more favourable. If the two polyenes interact so as to give an intermediate this is more likely to collapse to a small ring product than a large one, again for entropy reasons; just as, for example, the radical addition of olefins to dienes leads to more 2 + 2 than 2 + 4 addition. Cyclic polyenes are more likely to undergo cycloadditions leading to large rings but such cyclic systems as cyclo-octatetraenes, azacyclo-octatetraenes and oxepins exist in equilibrium with bicyclic valence tautomers (chapter 3) and frequently undergo cycloadditions via these tautomers (6.71). Tropones, some

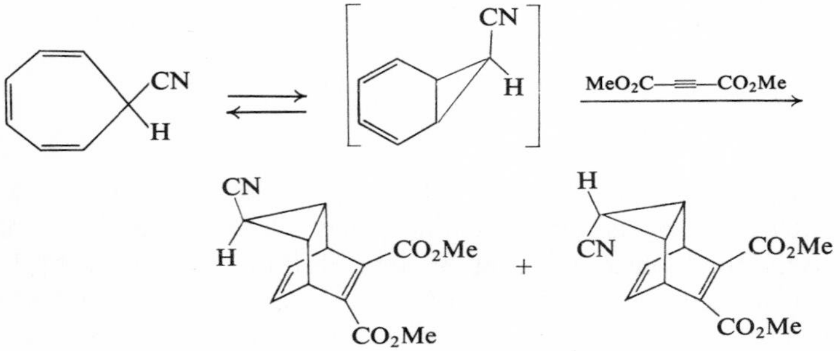

(6.71)

tropolones and 1-*H*-azepines do not usually undergo such valence isomerisations and have therefore more often been observed to behave as 6π electron components.

6 + 2 *Electron cycloadditions.* $_{\pi}6_s + _{\pi}2_s$ Cycloaddition is thermally disallowed and significantly 6 + 2 additions have only rarely been observed. The addition of nitrosobenzene to cycloheptatriene (72) and *N*-ethoxycarbonylazepine (73) is presumed to result from stepwise processes which are preferred over concerted 4 + 2 addition because of the highly polarised nitroso-group (6.72).[78]

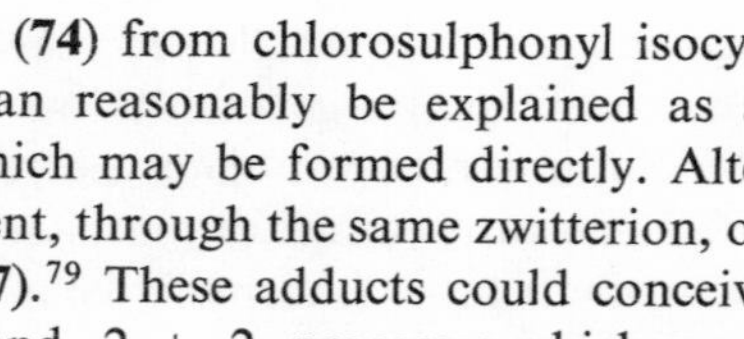

(72) X = CH$_2$
(73) X = NCO$_2$Et

(6.72)

The 6 + 2 adduct (74) from chlorosulphonyl isocyanate and cyclo-heptatriene (6.73) can reasonably be explained as arising from the intermediate (75) which may be formed directly. Alternatively it may arise by rearrangement, through the same zwitterion, of initially formed adducts (76) and (77).[79] These adducts could conceivably result from concerted $_{\pi}6_s + _{\pi}2_a$ and $_{\pi}2_s + _{\pi}2_a$ processes which are formally allowed for this cumulene.[80]

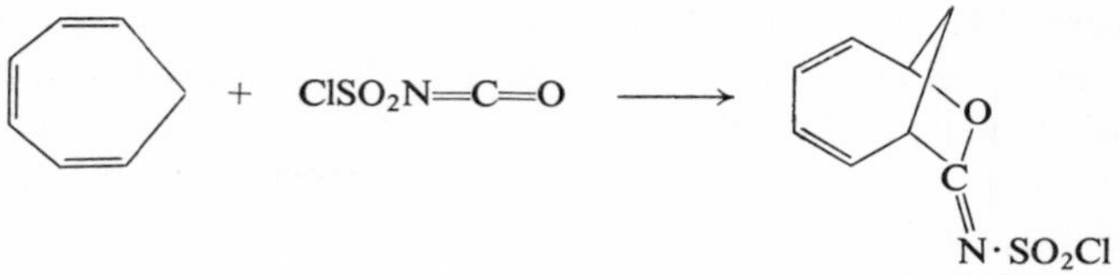

$$\text{(structure)} + \text{ClSO}_2\text{N}{=}\text{C}{=}\text{O} \longrightarrow \text{(structure)}$$

C
‖
N·SO₂Cl

(74)

(6.73)

ClSO₂

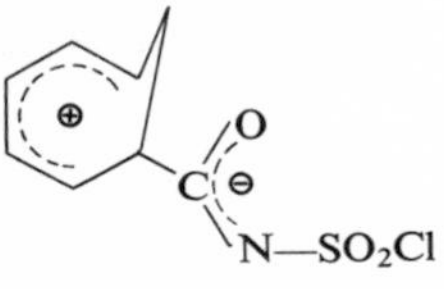

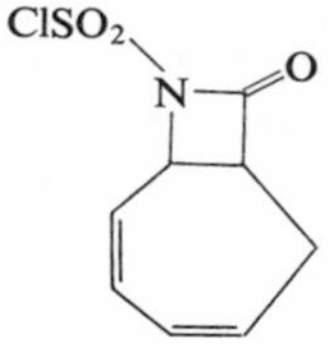

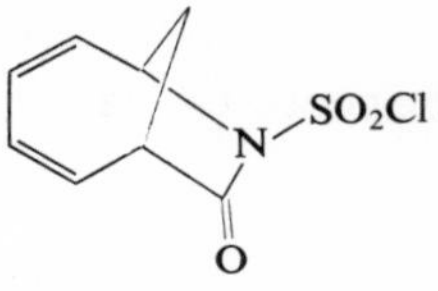

(75)   **(77)**   **(76)**

Sulphur dioxide adds readily to *cisoid*-hexatrienes by a linear cheletropic $_\pi6_a + {_\omega}2_s$ process (6.74–6.75). The reverse reaction is a linear cheletropic extrusion of sulphur dioxide with conrotatory twisting of the terminal methylenes.[81] With the cyclic triene (**78**) sulphur dioxide forms only the 1,4-adduct (**79**) (6.76). In this case antarafacial addition to the triene is geometrically impossible and the alternative non-linear $_\pi6_s + {_\omega}2_a$ process does not compete with the concerted $_\pi4_s + {_\omega}2_s$ linear cheletropic addition to a diene component. The extreme unfavourability of non-linear cheletropic eight-electron processes is illustrated by the fact that sulphur dioxide is eliminated 60000 times slower (at 180°C) from (**80**)

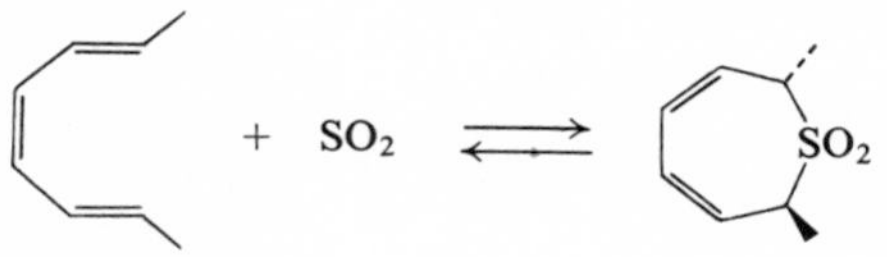

$$\text{(structure)} + \text{SO}_2 \rightleftharpoons \text{(structure)} \quad \text{SO}_2$$

(6.74)

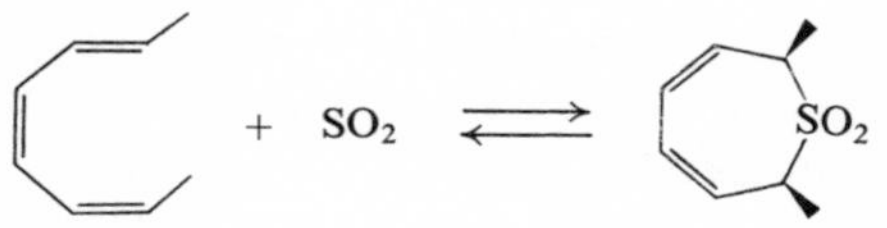

$$\text{(structure)} + \text{SO}_2 \rightleftharpoons \text{(structure)} \quad \text{SO}_2$$

(6.75)

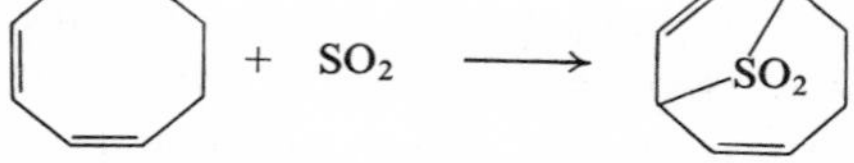

$$\text{(78)} + \text{SO}_2 \longrightarrow \text{(79)} \quad \text{SO}_2 \quad (6.76) \quad \text{(80)} \quad \text{SO}_2$$

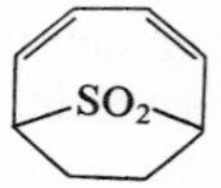

(78)   **(79)**   **(80)**

than from (**79**) (where an allowed retro-$_\pi 4_s + _\omega 2_s$ process is possible). The geometry of (**80**) makes conrotatory twisting of the two methylene groups (retro-$_\pi 6_a + _\omega 2_s$) impossible so that, if concerted, the elimination would have to be a non-linear cheletropic process (retro-$_\pi 6_s + _\omega 2_a$). This particular extrusion is therefore either stepwise or, if concerted, the non-linear cheletropic extrusion has an energy barrier of the same order of magnitude as a stepwise process.[82]

4π + 4π Cycloadditions. $_\pi 4_s + _\pi 4_s$ Cycloaddition should only be photochemically allowed. Several examples of photochemical 4 + 4 additions are known (e.g. 6.77) but as in other photochemical reactions the concerted nature is still in doubt.[1,83]

(6.77)

One example of a presumably stepwise thermal 4 + 4 cycloaddition is the dimerisation of diphenylisoindenone (**81**) illustrated in (6.78).[84] The highly reactive isoindenones normally function as dienes and can be trapped by dienophiles.[85]

(6.78)

(**81**)

8 + 2 Electron cycloadditions. $_\pi 8_s + _\pi 2_s$ Cycloadditions are thermally allowed but have rarely been observed.[1] The examples in fig. 6.31 may be concerted, but these are also cases where the dipolar intermediates would be particularly stabilised.[86]

The reaction of the cyclic ether (**82**) with 4-phenyltriazoline-3,5-dione at −78°C probably involves a concerted 8 + 2 cycloaddition.[87] The adduct (**83**) is not isolated but undergoes a disrotatory ring closure to give the observed product (**84**) on warming to about −25°C, (6.79). Evidence for this is that the diene (**84**) is extremely reactive towards the triazolinedione, and they form an adduct readily at −78°C. The absence

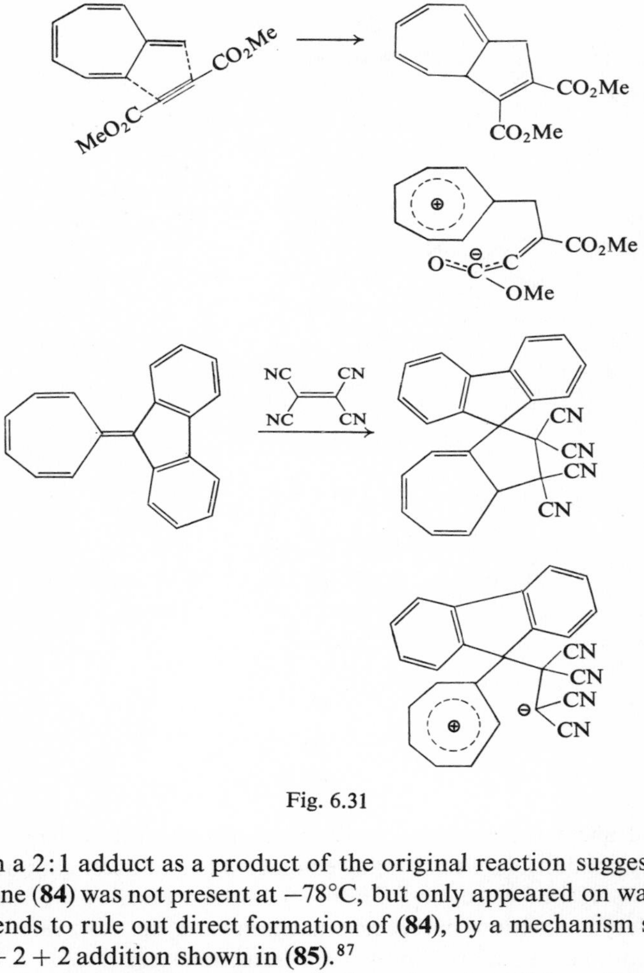

Fig. 6.31

of such a 2:1 adduct as a product of the original reaction suggests that the diene (**84**) was not present at −78°C, but only appeared on warming. This tends to rule out direct formation of (**84**), by a mechanism such as the 2 + 2 + 2 addition shown in (**85**).[87]

6π + 4π Cycloadditions. Thermal $_\pi6_s + _\pi4_s$ cycloadditions (6.80) were discovered after the prediction that they would be allowed. Several examples are now known.[1] The *exo*-orientation observed is in agreement with secondary orbital interaction (light dashes) which would disfavour the transition state for *endo*-addition (**86**).

N-Substituted 1*H*-azepines dimerise on heating to give 6 + 4 dimers (**87**). On further heating these rearrange by a diradical suprafacial 1,3-shift to give 6 + 6 dimers (**88**, 6.81). Originally, only the latter were

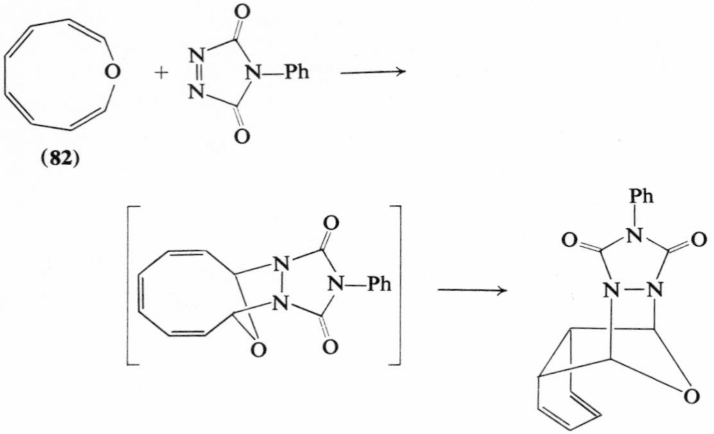

(82)

(6.79)

(83)   (84)

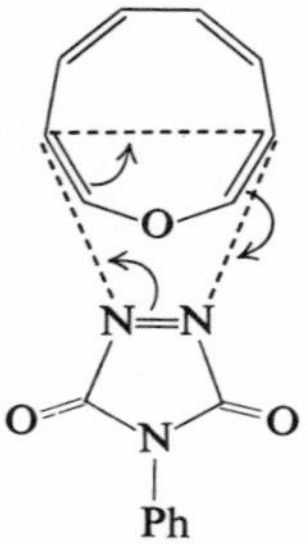

(85)

isolated under the reaction conditions and it was the recognition that
6 + 6 cycloadditions were disallowed that led to a reinvestigation and the
discovery of the 6 + 4 dimers. The 6 + 4 dimers are again *exo*; favourable
secondary interactions in the *exo*-transition state, fig. 6.32, reinforce the
unfavourable interactions in the *endo*-transition state.[88]

The π systems of azepines and tropones are such that they can also
undergo cycloadditions in which four or two π electrons are utilised.
For example, *N*-substituted azepines give 4 + 2 adducts with reactive
dienophiles such as tetracyanoethylene (6.82) and *N*-phenylmaleimide.
Significantly, *endo*-addition is observed in such cases, as for other Diels–
Alder reactions.   Reactive dienes such as isobenzofurans (6.83) and
2,5-dimethyl-3,4-diphenylcyclopentadienone add to the 4,5-bond of
azepines, again with predictable *endo*-stereospecificity.[89a]

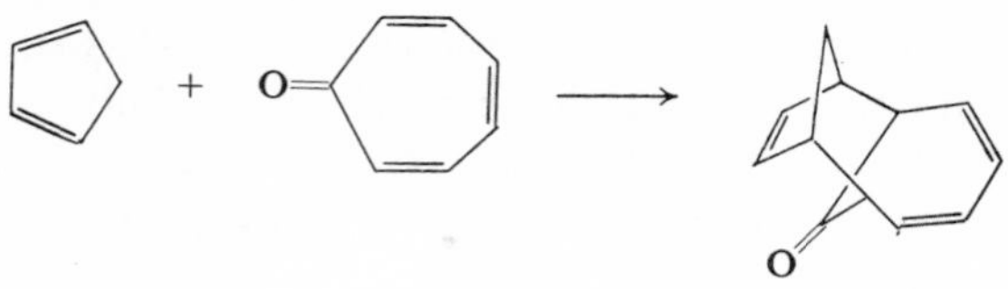

(6.80)

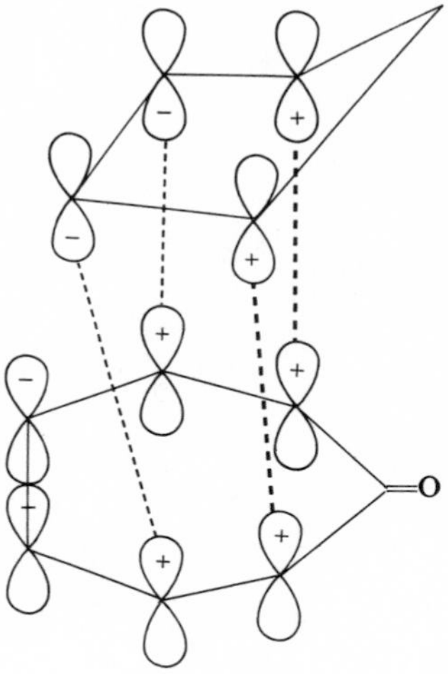

(86)

$6\pi + 4\pi$ Cycloadditions in which the 4π component is a 1,3-dipole have also been observed.[89b]

$6\pi + 6\pi$ *Cycloadditions.* Tropone gives a $6 + 6$ dimer (89) when irradiated (6.84). Photochemical $_\pi 6_s + _\pi 6_s$ cycloaddition is allowed, but there is no evidence that the dimer (89) is formed concertedly.[90]

Cycloadditions with more than twelve electrons. Cycloadditions involving fourteen electrons have not yet been observed. The heptafulvalene (90) is ideally twisted to undergo the allowed $_\pi 14_a + _\pi 2_s$ addition which has been observed with tetracyanoethylene (6.85).[1]

An attempt to observe $16\pi + 2\pi$ cycloaddition of acetylene dicarboxylic ester to the fulvene (91) gave the *cis* and *trans* isomers of the spiroadducts (92) by a stepwise process.[91]

Multicomponent cycloadditions. For entropy reasons, multicomponent additions involving more than six electrons are very unlikely unless more than one of the components is situated within the same molecule. The photochemically induced cycloaddition of but-2-yne to dihydrophthalic anhydride can be rationalised as a photochemically allowed $_\pi 2_s + {}_\pi 2_s + {}_\pi 2_s + {}_\pi 2_s$ addition, the orthogonal π bonds of the butyne being considered as separate 2π components (6.87). A stepwise mechanism involving an intermediate carbene (93) seems equally plausible, however.[92]

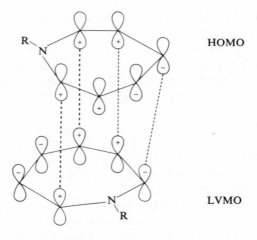

$$R = CO_2Et \qquad\qquad\qquad\qquad\qquad (6.81)$$

6 + 4 adduct (87)   6 + 6 adduct (88)

HOMO

LVMO

Fig. 6.32

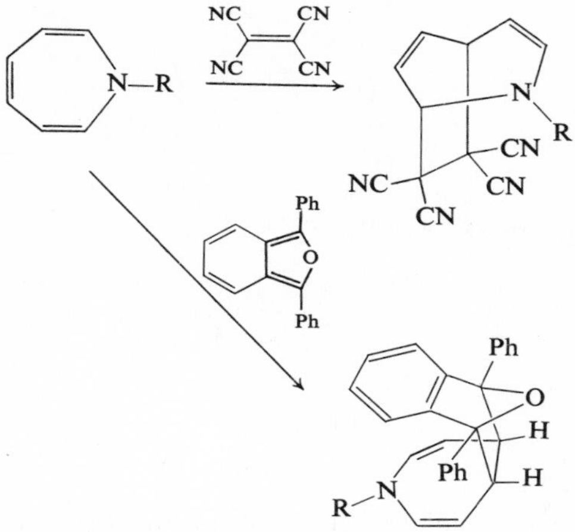

(6.82)

(6.83)

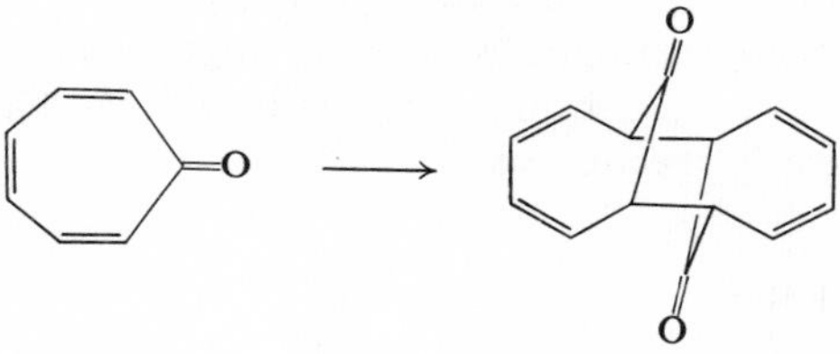

(6.84)

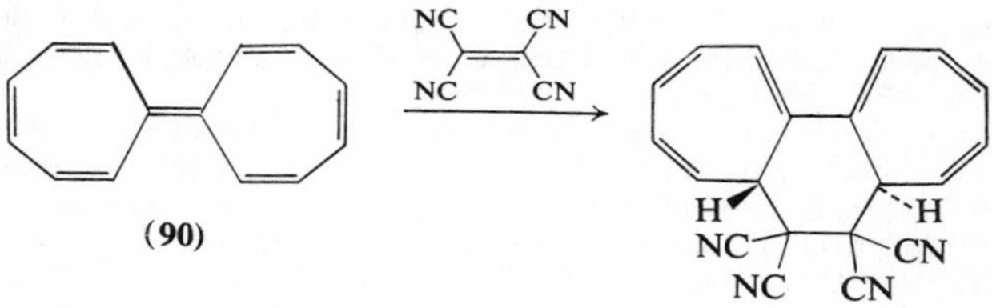

(**89**)

(6.85)

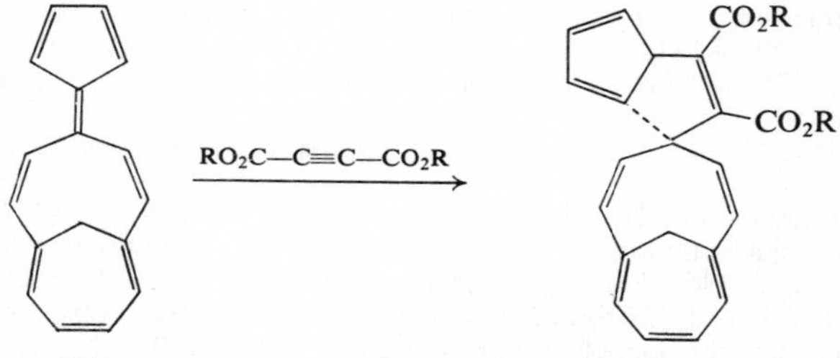

(**90**)

(6.86)

(**91**)

(**92**)

(6.87)

(93)

Multicomponent cycloadditions catalysed by metals are also possible. Examples are the nickel catalysed tetramerisation of acetylene to give cyclo-octatetraene (the Reppe synthesis), and catalysed cyclotrimerisation of acetylenes to give benzene derivatives.[93] Mechanisms for these reactions are largely speculative.

REFERENCES

1. R. B. Woodward and R. Hoffmann, *Angew. Chem. Int. Edn*, **8**, 781 (1969).
2. K. Kraft and G. Koltzenburg, *Tetrahedron Letters*, 1967, 4357, 4723.
3. K. Ziegler, H. Sauer, L. Bruns, H. Froitzheim-Kühlhorn, and J. Schneider, *Annalen*, **589**, 122 (1954); A. C. Cope, C. F. Howell, and A. Knowles, *J. Amer. Chem. Soc.*, **84**, 3190 (1962).
4. Reviews: (a) R. Gompper, *Angew. Chem. Int. Edn*, **8**, 312 (1969); (b) R. Huisgen, R. Grashey, and J. Sauer in *The Chemistry of Alkenes*, ed. S. Patai, Interscience, 1964, p. 741; (c) L. L. Müller and J. Hamer, *1,2-Cycloaddition Reactions*, Interscience, 1967; (d) P. D. Bartlett, *Quart. Reviews*, **24**, 473 (1970).
5. J. K. Williams, D. W. Wiley, and B. C. McKusick, *J. Amer. Chem. Soc.*, **84**, 2210 (1962).
6. M. E. Kuehne and L. Foley, *J. Org. Chem.*, **30**, 4280 (1965).
7. K. C. Brannock, A. Bell, R. D. Burpitt, and C. A. Kelly, *J. Org. Chem.*, **26**, 625 (1961); **29**, 801 (1964).
8. R. Huisgen in *Topics in Heterocyclic Chem.*, ed. R. N. Castle, Interscience, 1969, p. 223.
9. S. Proskow, H. E. Simmons, and T. L. Cairns, *J. Amer. Chem. Soc.*, **88**, 5254 (1966).
10. R. W. Hoffmann and H. Häuser, *Angew. Chem. Int. Edn*, **3**, 380 (1964).
11. (a) E. Koerner von Gustorf, D. V. White, J. Leitich, and D. Henneberg, *Tetrahedron Letters*, 1969, 3113; E. Koerner von Gustorf, D. V. White, B. Kim, D. Hess, and J. Leitich, *J. Org. Chem.*, **35**, 1155 (1970); (b) E. Koerner von Gustorf, D. V. White, and J. Leitich, *Tetrahedron Letters*, 1969, 3109.

12. K. C. Brannock, R. D. Burpitt, H. E. Davis, H. S. Pridgen, and J. G. Thweatt, *J. Org. Chem.*, **29**, 2579 (1964).
13. Reviews: (*a*) J. D. Roberts and C. M. Sharts, *Organic Reactions*, **12**, 1 (1962); (*b*) P. D. Bartlett, *Science*, **159**, 833 (1968).
14. L. K. Montgomery, K. Schueller, and P. D. Bartlett, *J. Amer. Chem. Soc.*, **86**, 622 (1964).
15. P. D. Bartlett, C. J. Dempster, L. K. Montgomery, K. E. Schueller, and G. E. H. Wallbillich, *J. Amer. Chem. Soc.*, **91**, 405 (1969); P. D. Bartlett and G. E. H. Wallbillich, *ibid.*, **91**, 409 (1969).
16. J. S. Swenton and P. D. Bartlett, *J. Amer. Chem. Soc.*, **90**, 2056 (1968).
17. P. D. Bartlett, G. E. H. Wallbillich, A. S. Wingrove, J. S. Swenton, L. K. Montgomery, and B. D. Kramer, *J. Amer. Chem. Soc.*, **90**, 2049 (1968).
18. P. D. Bartlett and R. Wheland, *J. Amer. Chem. Soc.*, **92**, 3822 (1970).
19. J. C. Little, *J. Amer. Chem. Soc.*, **87**, 4020 (1965).
20. P. D. Bartlett and K. E. Schueller, *J. Amer. Chem. Soc.*, **90**, 6077 (1968).
21. P. D. Bartlett and K. E. Schueller, *J. Amer. Chem. Soc.*, **90**, 6071 (1968).
22. For a comprehensive review of heterocumulene cycloadditions, see H. Ulrich, *Cycloaddition Reactions of Heterocumulenes*, Academic Press, 1968. For reviews of allene and other cumulene cycloadditions, see A. Fischer in *The Chemistry of Alkenes*, ed. S. Patai, Interscience, 1964, p. 1025, and D. R. Taylor, *Chem. Reviews*, **67**, 317 (1967).
23. R. Montaigne and L. Ghosez, *Angew. Chem. Int. Edn*, **7**, 221 (1968).
24. R. Huisgen and P. Otto, *Tetrahedron Letters*, 1968, 4491.
25. R. Huisgen, L. A. Feiler, and P. Otto, *Chem. Berichte*, **102**, 3405, 3444 (1969); R. Huisgen, L. A. Feiler, and G. Binsch, *ibid.*, 3460; R. Huisgen and L. A. Feiler, *ibid.*, 3391, 3428; R. Huisgen and P. Otto, *ibid.*, 3475.
26. J. E. Baldwin and J. A. Kapecki, *J. Amer. Chem. Soc.*, **92**, 4868, 4874 (1970).
27. Reference 26 gives a slightly different picture of the secondary interaction which is claimed to explain the observed isotope effects. See also H. U. Wagner and R. Gompper, *Tetrahedron Letters*, 1970, 2819, for an alternative two-stage dipolar mechanism for keten cycloaddition.
28. P. R. Brook, J. M. Harrison, and A. J. Duke, *Chem. Comm.*, 589 (1970); P. R. Brook, A. J. Duke, and J. R. C. Duke, *ibid.*, 574 (1970); W. T. Brady and R. Roe, *J. Amer. Chem. Soc.*, **92**, 4618 (1970); W. T. Brady and E. F. Hoff, *J. Org. Chem.*, **35**, 3733 (1970).
29. R. Huisgen and P. Otto, *J. Amer. Chem. Soc.*, **90**, 5342 (1968).
30. P. Otto, L. A. Feiler, and R. Huisgen, *Angew. Chem. Int. Edn*, **7**, 737 (1968).
31. R. Huisgen and P. Otto, *J. Amer. Chem. Soc.*, **91**, 5922 (1969).
32. (*a*) R. Huisgen, B. A. Davies, and M. Morikawa, *Angew. Chem. Int. Edn*, **7**, 826 (1968); (*b*) H. B. Kagan and J. C. Luche, *Tetrahedron Letters*, 1968, 3093; (*c*) A. Gomes and M. M. Joullié, *Chem. Comm.*, 1967, 935.
33. W. T. Brady and E. D. Dorsey, *Chem. Comm.*, 1968, 1638.
34. W. Neumann and P. Fischer, *Angew. Chem. Int. Edn*, **1**, 621 (1962).
35. H. N. Cripps, J. K. Williams, and W. H. Sharkey, *J. Amer. Chem. Soc.*, **81**, 2723 (1959).
36. E. F. Kiefer and M. Y. Okamura, *J. Amer. Chem. Soc.*, **90**, 4187 (1968), and references therein.
37. J. E. Baldwin and U. V. Roy, *Chem. Comm.*, 1969, 1225.
38. W. R. Moore, R. D. Bach, and T. M. Ozretich, *J. Amer. Chem. Soc.*, **91**, 5918 (1969).
39. W. R. Dolbier and S.-H. Dai, *J. Amer. Chem. Soc.*, **92**, 1774 (1970), and references therein.

40. For a review of acid catalysed cyclodimerisations via vinyl cation intermediates see K. Griesbaum, *Angew. Chem. Int. Edn*, **8**, 933 (1969).
41. K. Griesbaum, W. Nägele, and G. G. Wanless, *J. Amer. Chem. Soc.*, **87**, 3151 (1965).
42. R. Criegee and A. Moschel, *Chem. Berichte*, **92**, 2181 (1959).
43. Reviews: H. M. Frey and R. Walsh, *Chem. Reviews*, **69**, 103 (1969); *Advances Phys. Org. Chem.*, **4**, 170 (1966).
44. H. R. Gerberich and W. D. Walters, *J. Amer. Chem. Soc.*, **83**, 3935, 4884 (1961).
45. A. T. Cocks, H. M. Frey, and I. D. R. Stevens, *Chem. Comm.*, 1969, 458.
46. J. E. Baldwin and P. W. Ford, *J. Amer. Chem. Soc.*, **91**, 7192 (1969).
47. L. A. Paquette and J. A. Schwartz, *J. Amer. Chem. Soc.*, **92**, 3215 (1970).
48. N. Rieber, J. Alberts, J. A. Lipsky, and D. M. Lemal, *J. Amer. Chem. Soc.*, **91**, 5668 (1969).
49. G. L. Closs and P. E. Pfeffer, *J. Amer. Chem. Soc.*, **90**, 2452 (1968).
50. D. S. Noyce and E. H. Banitt, *J. Org. Chem.*, **31**, 4043 (1966); M. U. S. Sultanbawa, *Tetrahedron Letters*, 1968, 4569.
51. L. A. Paquette, M. J. Wyvratt and G. R. Allen, *J. Amer. Chem. Soc.*, **92**, 1763 (1970).
52. E. H. White, J. Wiecko, and C. C. Wei, *J. Amer. Chem. Soc.*, **92**, 2167 (1970); H. Güsten and E. F. Ullman, *Chem. Comm.*, 1970, 28.
53. F. McCapra, *Chem. Comm.*, 1968, 155.
54. Review: B. P. Stark and A. J. Duke, *Extrusion Reactions*, Pergamon Press, 1967.
55. Reviews: (a) W. Kirmse, *Carbene Chemistry*, Academic Press, 1971; (b) W. Lwowski, *Nitrenes*, Interscience, 1970; (c) T. L. Gilchrist and C. W. Rees, *Carbenes, Nitrenes, and Arynes*, Nelson, 1969; (d) D. Bethell, *Advances Phys. Org. Chem.*, **7**, 153 (1969).
56. R. Hoffmann, *J. Amer. Chem. Soc.*, **90**, 1475 (1968).
57. P. S. Skell and R. C. Woodworth, *J. Amer. Chem. Soc.*, **78**, 4496 (1956).
58. N. P. Neureiter, *J. Amer. Chem. Soc.*, **88**, 558 (1966); L. A. Paquette, *Accounts Chem. Research*, **1**, 209 (1968).
59. F. G. Bordwell, J. M. Williams, E. B. Hoyt, and B. B. Jarvis, *J. Amer. Chem. Soc.*, **90**, 429 (1968); D. C. Dittmer, G. C. Levy, and G. E. Kuhlmann, *ibid.*, **91**, 2097 (1969).
60. G. E. Hartzell and J. N. Paige, *J. Org. Chem.*, **32**, 459 (1967).
61. R. D. Clark and G. K. Helmkamp, *J. Org. Chem.*, **29**, 1316 (1964).
62. J. P. Freeman and W. H. Graham, *J. Amer. Chem. Soc.*, **89**, 1761 (1967).
63. L. A. Carpino and R. K. Kirkley, *J. Amer. Chem. Soc.*, **92**, 1784 (1970).
64. H. W. Heine, J. D. Myers, and E. T. Peltzer, *Angew. Chem. Int. Edn*, **9**, 374 (1970).
65. Reviews: (a) O. L. Chapman and G. Lenz in *Organic Photochemistry*, vol 1, Arnold, 1967, p. 283; (b) R. N. Warrener and J. B. Bremner, *Reviews Pure Appl. Chem.*, **16**, 117 (1966); (c) W. L. Dilling, *Chem. Reviews*, **66**, 373 (1966); (d) A. A. Lamola and N. J. Turro, *Technique of Organic Chemistry* **14**, Interscience, 1969; see also J. L. Vollmer and K. L. Servis, *J. Chem. Ed.*, **47**, 491 (1970).
66. D. R. Arnold and V. Y. Abraitys, *Chem. Comm.*, 1967, 1053.
67. H. Yamazaki and R. J. Cvetanović, *J. Amer. Chem. Soc.*, **91**, 520 (1969).
68. J. Saltiel and L. S. N. Lim, *J. Amer. Chem. Soc.*, **91**, 5404 (1969).
69. C. D. DeBoer and R. H. Schlessinger, *J. Amer. Chem. Soc.*, **90**, 803 (1968).
70. P. E. Eaton, *Accounts Chem. Research*, **1**, 50 (1968).
71. E. J. Corey, R. B. Mitra, and H. Uda, *J. Amer. Chem. Soc.*, **86**, 485 (1964); E. J. Corey and S. Nozoe, *ibid.*, **86**, 1652 (1964).
72. R. A. Deering and R. B. Setlow, *Biochim. Biophys. Acta*, **68**, 526 (1963).

73. (a) F. D. Mango, *Advances Catalysis*, **20**, 291 (1969); (b) G. N. Schrauzer, *ibid.*, **18**, 373 (1968); (c) R. Pettit, H. Sugahara, J. Wristers, and W. Merk, *Disc. Faraday Soc.*, **47**, 71 (1969).
74. H. Hogeveen and H. C. Volger, *J. Amer. Chem. Soc.*, **89**, 2486 (1967).
75. N. Calderon, E. A. Ofstead, J. P. Ward, W. A. Judy, and K. W. Scott, *J. Amer. Chem. Soc.*, **90**, 4133 (1968). See also ref. 73 (a), (c).
76. R. Wolovsky, *J. Amer. Chem. Soc.*, **92**, 2132 (1970); D. A. Ben-Efraim, C. Batich, and E. Wasserman, *ibid.*, **92**, 2133 (1970).
77. L. Cassar, P. E. Eaton, and J. Halpern, *J. Amer. Chem. Soc.*, **92**, 3515 (1970).
78. P. Burns and W. A. Waters, *J. Chem. Soc.* (C), 1969, 27; W. S. Murphy and J. R. McCarthy, *Chem. Comm.*, 1968, 1155.
79. L. A. Paquette, S. Kirschner, and J. R. Malpass, *J. Amer. Chem. Soc.*, **91**, 3970 (1969).
80. See E. J. Moriconi, C. F. Hummel, and J. F. Kelly, *Tetrahedron Letters*, 1969, 5325, and references therein, for possible examples of concerted additions of this cumulene.
81. W. L. Mock, *J. Amer. Chem. Soc.*, **89**, 1281 (1967); *ibid.*, **91**, 5682 (1969).
82. W. L. Mock, *J. Amer. Chem. Soc.*, **92**, 3807 (1970).
83. For the specific example shown see L. A. Paquette and G. Slomp, *J. Amer. Chem. Soc.*, **85**, 765 (1963).
84. J. M. Holland and D. W. Jones, *Chem. Comm.*, 1969, 587.
85. K. Blatt and R. W. Hoffmann, *Angew. Chem. Int. Edn*, **8**, 606 (1969).
86. W. von E. Doering and D. W. Wiley, *Tetrahedron*, **11**, 183 (1960); H. Prinzbach, D. Seip, L. Knothe, and W. Faisst, *Annalen*, **698**, 34 (1966).
87. A. G. Anastassiou and R. P. Cellura, *Chem. Comm.*, 1970, 484.
88. I. C. Paul, S. M. Johnson, J. H. Barrett, and L. A. Paquette, *Chem. Comm.*, 1969, 6; L. A. Paquette, J. H. Barrett, and D. E. Kuhla, *J. Amer. Chem. Soc.*, **91**, 3616 (1969).
89. (a) L. A. Paquette, D. E. Kuhla, J. H. Barrett, and L. M. Leichter, *J. Org. Chem.*, **34**, 2888 (1969); J. R. Wiseman and B. P. Chong, *Tetrahedron Letters*, 1969, 1619; (b) K. N. Houk and C. R. Watts, *Tetrahedron Letters*, 1970, 4025; K. N. Houk and L. J. Luskus, *ibid.*, 4029.
90. T. Mukai, T. Tezuka, and Y. Akasaki, *J. Amer. Chem. Soc.*, **88**, 5025 (1966).
91. H. Prinzbach, L. Knothe, and A. Dieffenbacher, *Tetrahedron Letters*, 1969, 2093.
92. R. Huisgen, *Angew. Chem. Int. Edn*, **7**, 321 (1968), and references therein.
93. Review: W. Reppe, N. Kutepov and A. Magin, *Angew. Chem. Int. Edn*, **8**, 727 (1969).

7 Sigmatropic rearrangements

Uncatalysed thermal rearrangements occurring intramolecularly and involving a six-membered cyclic transition state are very common, and include an enormous variety of structural types. Some of these, such as the Cope and Claisen rearrangements and 1,5-hydrogen shifts in conjugated dienes, have been known for many years; others have only recently been recognised as being of this type. The development of the theory of concerted reactions has led to the realisation that they represent only one of many possible classes of similar rearrangements, some thermally and some photochemically induced, which could occur in anions and cations as well as in neutral molecules. The unifying features of all these reactions are that they are concerted, uncatalysed, and involve a bond migration through a cyclic transition state in which an atom or group is simultaneously joined to both termini of a π electron system. Woodward and Hoffmann have given the name *sigmatropic rearrangements* to such reactions, the adjective 'sigmatropic' indicating movement of a sigma bond.

7.1. Nomenclature. There is a formal system of nomenclature for sigmatropic rearrangements which is widely used. Consider the Cope rearrangement of 1,5-hexadiene (fig. 7.1): in this reaction a σ bond is broken and one is made, and two π bonds migrate. The termini of the σ bond have moved to carbon atoms 3 and 3′, according to the numbering shown, with the original termini at carbons 1 and 1′. This change is then defined as a sigmatropic reaction of order [3, 3], both termini having moved to the *third* carbon atom along the π system. In general, a change of order $[i, j]$ involves a movement of a σ bond to a new position where its termini are i-1 and j-1 atoms removed from the original position.

A few other examples will illustrate the system. The feasibility of the reactions will be discussed later; the examples are given to show the use of the numbering system. In each case the numbering starts at the original termini of the σ bond.

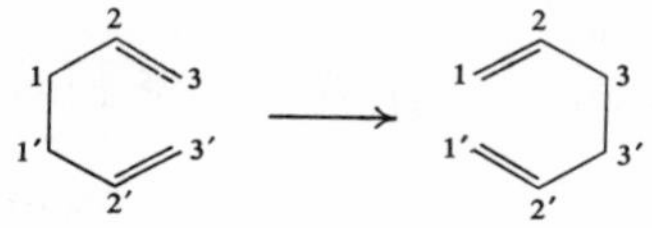

Fig. 7.1

The ylide rearrangement (*a*) is a sigmatropic change of order [2, 3], since the new σ bond is one and two atoms away from the original position.

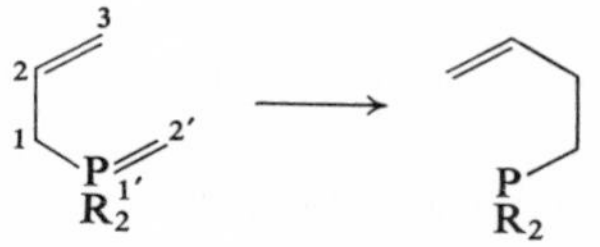

(*a*)

The migration of hydrogen through an allyl system (*b*) is a rearrangement of order [1, 3] because the univalent hydrogen has moved from C-1 to C-3 ($j = 3$); the other end of the σ bond is still attached to hydrogen

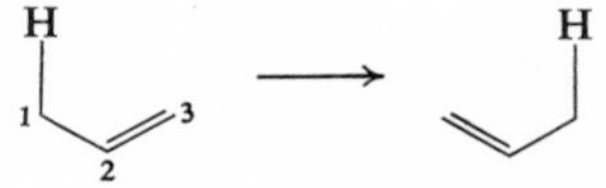

(*b*)

and so $i = 1$. Similarly, any migration in which an atom or a group moves unchanged through a π system will be a rearrangement of order [1, *j*].

The norcaradiene rearrangement (*c*) is a [1, 5] shift, the alkyl group

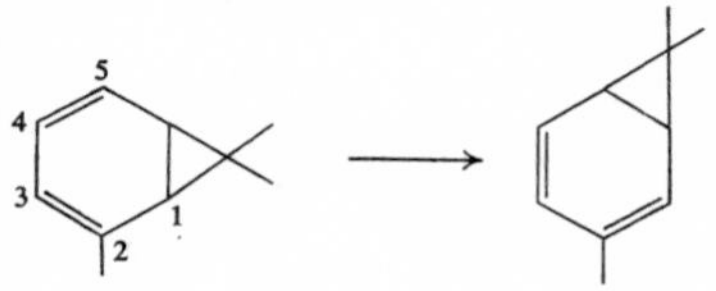

(*c*)

at C-1 migrating to C-5. The numbering must go through the π system, even though an alternative numbering through the adjacent tetrahedral carbon would give a smaller figure.

A simple rule for determining the order [*i, j*] is as follows: count the number of atoms in each of the fragments formally produced by breaking the migrating σ bond. This gives *i* and *j* directly.

7.2. Hydrogen migrations

Selection rules. Consider the $[1, j]$ shift of a hydrogen atom between the ends of a polyene.

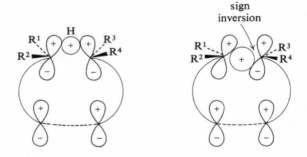

In a cyclic transition state, the bonding hydrogen orbital must overlap simultaneously with orbitals on both the terminal carbon atoms. These carbon orbitals must also overlap with the other p-orbitals of the polyene chain. The orbital on C-1 eventually becomes a p-orbital of the polyene while that on C-j becomes the sp^3-bonding orbital of the new C—H bond.

There are two stereochemically distinct ways in which the overlap can take place in the transition state. The spherically symmetrical hydrogen orbital can overlap with the p-lobes on the same side of the π system (*suprafacial* overlap) or on opposite sides (*antarafacial* overlap). The geometry of the two transition states is different (fig. 7.2); the suprafacial migration is characterised by a plane of symmetry, and the antarafacial migration by a twofold axis.

Selection rules for such migrations can most easily be derived by applying the 'aromatic transition state' concept. The signs of the lobes in fig. 7.2 are put in according to the method described in chapter 2; that is, with the maximum number of adjacent overlapping lobes having the same sign. The transition state can then be classed as being of the Hückel

Fig. 7.2. (Left) Suprafacial overlap, Hückel type; (right) Antarafacial overlap, Möbius type.

type (no sign inversions in the cycle) or of the Möbius type (one sign inversion). The advantage of the method for these systems is that it does not require a knowledge of the symmetries of the HOMO and LVMO of the reactants. Selection rules for hydrogen migrations can then be deduced.

The Hückel type transition state is favoured when 2, 6, ..., $(4n + 2)$ electrons participate, and the Möbius type when 4, 8, ..., $4n$ electrons participate. Thus *it is the number of electrons, not the number of atoms, which determines the selection rules.* This is particularly important to

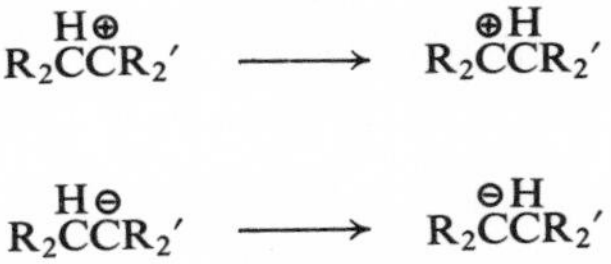

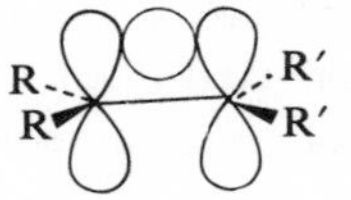

$$\overset{H\oplus}{R_2CCR_2'} \longrightarrow \overset{\oplus H}{R_2CCR_2'}$$

$$\overset{H\ominus}{R_2CCR_2'} \longrightarrow \overset{\ominus H}{R_2CCR_2'}$$

Fig. 7.3. Transition state for suprafacial [1, 2] shift.

bear in mind when considering migrations in charged polyenes. For example, a [1, 2] hydrogen shift in a carbonium ion is favoured as a suprafacial process, since two electrons (those of the σ bond) participate; but a [1, 2] hydrogen shift in a carbanion is not favoured as a suprafacial process, since four electrons (two from the anion and two from the σ bond) participate (fig. 7.3). A general formulation of the selection rules for thermal sigmatropic shifts of hydrogen, based on the number of participating electrons, is given in table 7.1.

TABLE 7.1. *Selection rules for sigmatropic hydrogen migrations*

Number of electrons	Neutral	Polyene Cation	Anion	Thermally allowed migration
2	—	[1, 2]	—	suprafacial
4	[1, 3]	[1, 4]	[1, 2]	antarafacial
6	[1, 5]	[1, 6]	[1, 4]	suprafacial
$4n$	$[1, (4n - 1)]$	$[1, 4n]$	$[1, (4n - 2)]$	antarafacial
$4n + 2$	$[1, (4n + 1)]$	$[1, (4n + 2)]$	$[1, 4n]$	suprafacial

Other approaches can be used to derive these selection rules. Correlation diagrams are difficult to construct because neither the starting materials nor the products of sigmatropic migrations have the requisite symmetry; but they have been used. Similarly the frontier orbital method can be applied if certain assumptions are made about the transition state. For example, the transition states of $[1, j]$ hydrogen migrations can be regarded as the interaction of a hydrogen atom with a polyene radical. The part-filled HOMO of the radical has the symmetry shown; note that here (in contrast to fig. 7.2) the + and − signs indicate the symmetry of the HOMO.

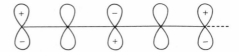

The termini of the system are in phase for polyenes of five or nine atoms, and out of phase for three or seven atoms (see chapter 2). Thus a bonding interaction with the part-filled hydrogen s-orbital will involve suprafacial overlap for the five- or nine-membered carbon chains, and antarafacial overlap for the three- or seven-membered chains (fig. 7.4). These can be regarded as the HOMOs of the transition states.

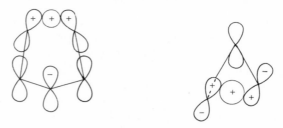

Fig. 7.4. (Left) [1, 5] shift (suprafacial); (right) [1, 3] shift (antarafacial).

Similar arguments can be used to derive the selection rules for charged polyenes. The aromatic transition state concept seems preferable here, however, since it avoids the necessity of working out symmetries of HOMOs.

Examples. The selection rules predict that suprafacial [1, 5] hydrogen shifts in neutral polyenes are thermally allowed, whereas [1, 3] and [1, 7] shifts must go by an antarafacial process. This implies that thermal [1, 3] shifts of hydrogen are unlikely to be observed, since the transition state for antarafacial migration would be very strained and difficult to attain. On the other hand, [1, 5] shifts should be facile. The experimental observations fit in with this pattern: whereas concerted uncatalysed [1, 3] hydrogen shifts have not been established, [1, 5] shifts in dienes (fig. 7.5) are well known, and there is evidence to support the view that they are concerted reactions.[1]

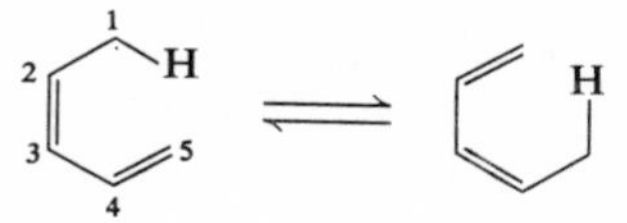

Fig. 7.5. [1, 5] Hydrogen shift.

In acyclic dienes, the [1, 5] shifts have activation energies of about 32 kcal mol^{-1} (134 kJ mol^{-1}) and they show large deuterium isotope effects (k_H/k_D is about 5 at 200°C in *cis*-1,3-pentadiene). The stereospecific suprafacial nature of the migration has been demonstrated with the diene (**1**).[2] As the diagram (equation 7.1) shows, the optically active starting material gave the two isomers expected from a suprafacial [1, 5] shift, but gave neither of the isomers that would result from an antarafacial migration.

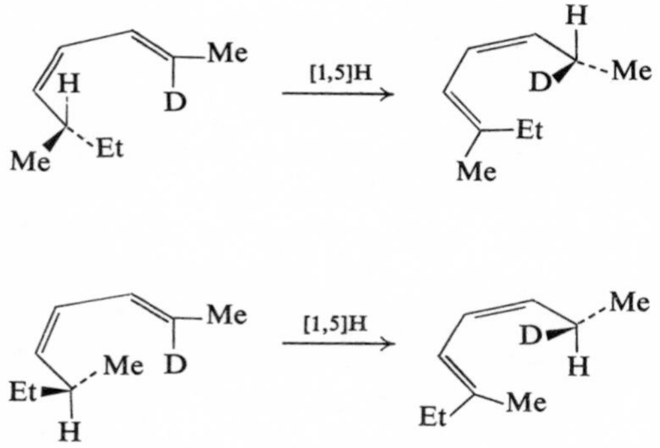

(7.1)

(**1**)

Thermal [1, 5] hydrogen shifts seem to be quite general in acyclic dienes which can achieve the necessary six-membered transition state geometry. In cyclic dienes, the rearrangement may go just as readily, if the geometry of the systems is similar to that required for the [1, 5] shift. The rearrangement has been observed in five-membered and larger rings. For example, 5-deuteriocyclo-octadiene (2) rearranged at 150°C by a succession of [1, 5] hydrogen shifts (7.2). When the system had reached equilibrium, the deuterium label was on all possible positions in the molecule. Note that (2) can also undergo a [1, 5] deuterium shift, but that this rearrangement is degenerate; that is, the product has the same structure as the starting material.

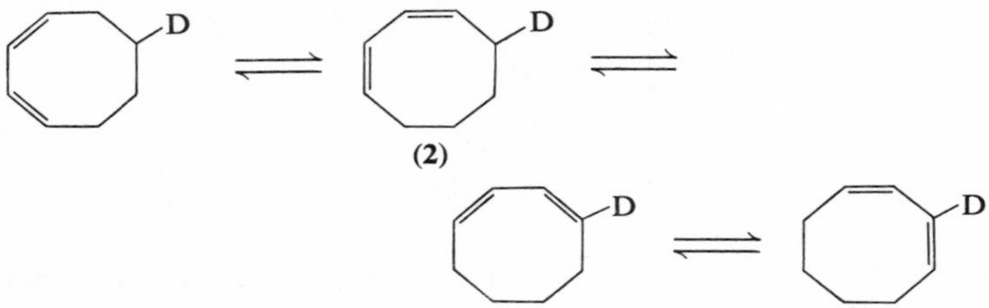

(2)

(7.2)

When 1-deuterioindene (3) was heated at 200°C, the deuterium became 'scrambled' over all three non-benzenoid carbons (7.3). The location of the deuterium at the 2-position indicates that the system prefers to rearrange by successive [1, 5] shifts rather than by [1, 3] shifts, even though [1, 5] shifts involve the intermediacy of isoindenes.

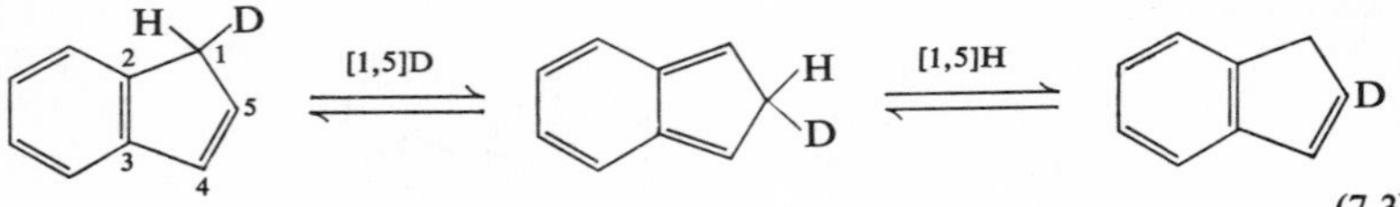

(3)   (7.3)

The simplest analogy for the reaction outside all-carbon systems appears to be the rearrangement of the unsaturated enol (5) formed by tautomerisation of the β,γ-unsaturated carbonyl compound (4). The overall result of the rearrangement is that the double bond is brought into conjugation with the carbonyl group (7.4). Such a rearrangement might be expected to be thermally induced, or acid catalysed, the acid promoting the enolisation of the carbonyl compound. Thus, the [1, 5] shift in the enol is a possible mechanism, though not the only one, for the movement of a double bond into conjugation with a carbonyl group.

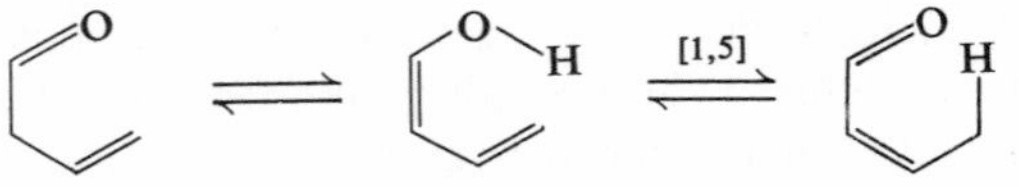

(4)   (5)   (7.4)

One reaction which probably does involve this mechanism is the thermal equilibration of α,β- and β,γ-unsaturated esters; for example (6) ⇌ (7), (7.5).[3] The activation parameters, measured for the reverse reaction, are similar to those for other [1, 5] shifts, with a low activation energy and a negative entropy of activation.

(6)   (7)   (7.5)

It is a general feature of pericyclic reactions that a cyclopropane ring can often participate in place of a double bond. Indeed, the olefin-like character of cyclopropane derivatives is well known; for example, in their ability to transmit conjugation. Various pictures of the bonding in cyclopropane have been constructed to explain this behaviour.[4] In the Walsh model (8), each carbon has an sp²-orbital directed towards the centre of the ring, and a p-orbital the lobes of which overlap with the p-orbitals on the adjacent carbons. These p-lobes can overlap with other p-lobes in a conjugated π system if the cyclopropane ring is suitably orientated; that is, with the plane of the ring parallel to the plane in which the rest of the π system lies.

(8)

Thus, a well documented extension of the [1, 5] shift is in systems where one of the double bonds is replaced by a cyclopropane ring. Such reactions have been called *homodienyl* [1, 5] hydrogen shifts (fig. 7.6).[1] The transition state for such rearrangements will be very similar to that for dienyl shifts, and the selection rules will similarly apply.

Such reactions are known both in acyclic and in cyclic systems, and appear to be mechanistically very similar to the dienyl shifts, with activation parameters of the same order. An example of the rearrangement in a cyclic system is the thermal equilibration of bicyclo[6,1,0]non-2-ene (9) and *cis,cis*-1,4-cyclononadiene at 150–170°C (7.6).

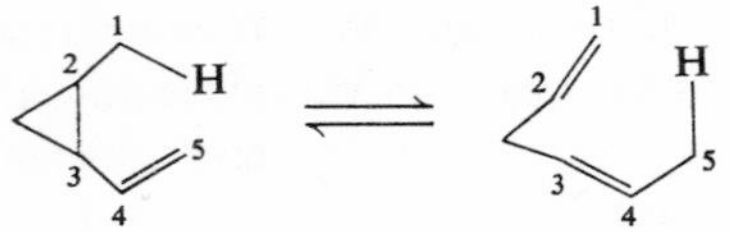

Fig. 7.6. Homodienyl [1, 5] shift.

The geometry of the transition state requires that the migrating hydrogen and the methylene group of the cyclopropane are *anti*, with the plane of the ring parallel to that of the π bond. This constraint makes

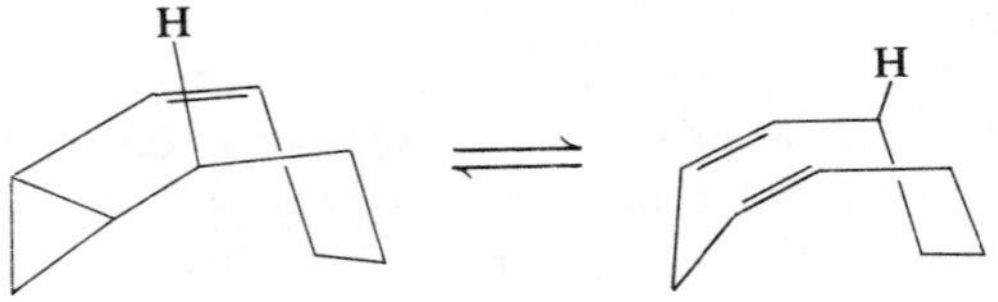

(7.6)

(9)

the transition state increasingly difficult to achieve as the ring size is reduced. In rings of smaller than seven atoms the activation energy is too great for the reaction to be a significant one.

The oxygen analogue of the reaction, the rearrangement of a cyclopropyl ketone, is also known (7.7).[5] This reaction provides a possible route for the isomerisation of γ,δ-unsaturated carbonyl compounds (7.8). In aromatic substrates the reaction is called the abnormal Claisen rearrangement; it is described with the Claisen rearrangement in § 7.4.

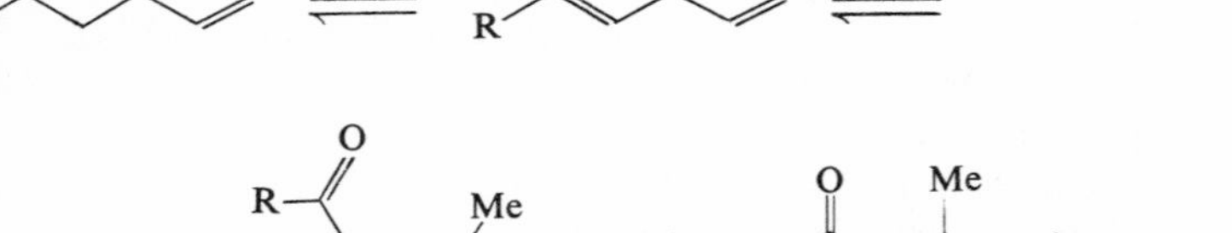

(7.7)

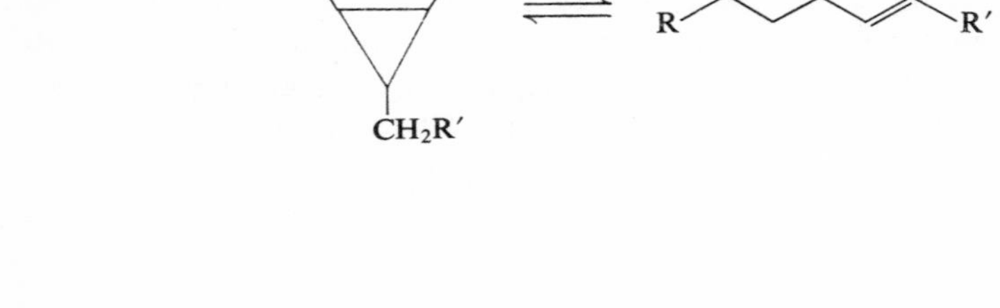

(7.8)

There is no hard and fast dividing line between [1, 5] shifts in which one of the participating bonds is a π bond or a 'π type' bond, as in the systems just described ($n = 0$ or 1 in structure **10**), and those in which it is a σ bond ($n > 1$ in **10**; or structure **11**). Mechanistically, all such reactions are closely related. The analogous reaction of σ systems such as (**11**) is the retro-ene reaction (§ 7.8).

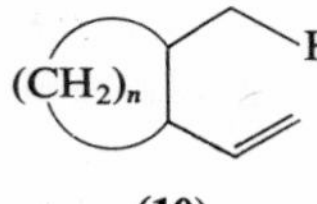

(**10**) 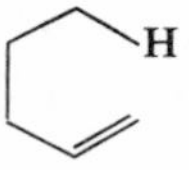   (**11**)

Antarafacial [1, 7] shifts are also thermally allowed, and the geometry of the transition state is not inaccessible (fig. 7.7).

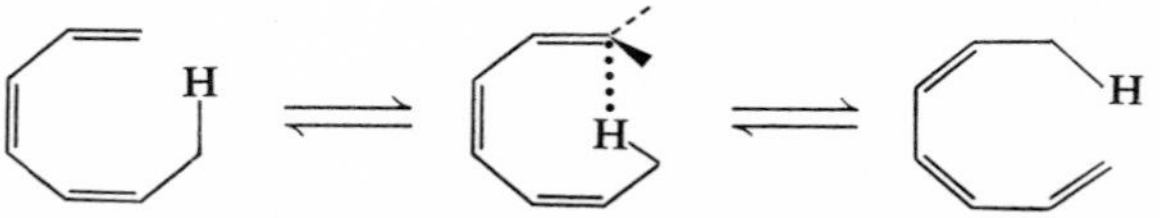

Fig. 7.7. Antarafacial [1, 7] shift.

A few reactions are known which probably do involve [1, 7] shifts of this type; one is the thermal interconversion of vitamin D (**12**) and precalciferol (**13**). In this system, the [1, 7] shift takes place very readily, and it may be that the apparent rarity of [1, 7] shifts is simply a reflection of the small number of suitable systems that have been investigated.[6]

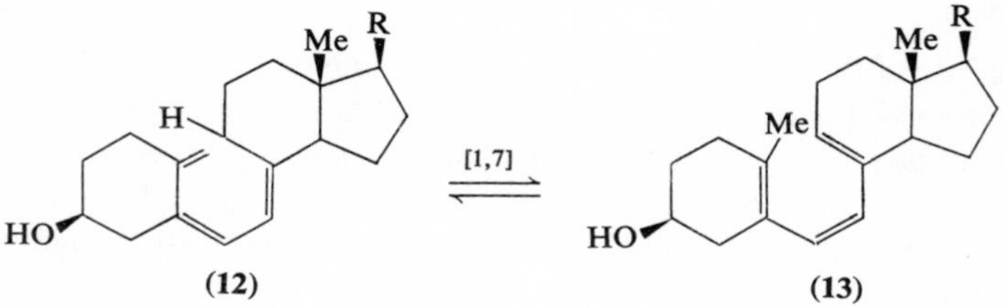

(**12**)   (**13**)

In charged systems, suprafacial [1, 2] shifts in cations are allowed. The 1,2-hydrogen or alkyl shift is indeed well known; for example, in the n-propyl to isopropyl cation rearrangement. The reaction will be discussed further in the next section.

7.3. Migrations of atoms other than hydrogen

Selection rules. The $[1, j]$ shift of a group other than hydrogen will obey the same selection rules as the hydrogen migrations if the transition state involves the same type of bonding, There are alternative types of bonding which might operate with groups other than hydrogen, however. Consider the case of the suprafacial [1, 3] migration of an alkyl group (fig. 7.8). The transition state which involves simultaneous bonding to the same lobe of the orbital of the alkyl group has the same form as that for hydrogen migration, and so will be thermally forbidden. On the other

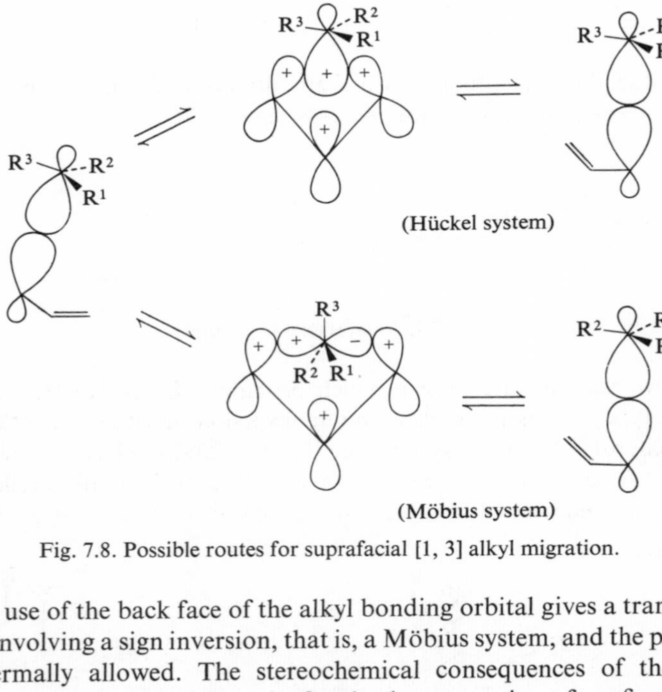

(Hückel system)

(Möbius system)

Fig. 7.8. Possible routes for suprafacial [1, 3] alkyl migration.

hand, use of the back face of the alkyl bonding orbital gives a transition state involving a sign inversion, that is, a Möbius system, and the process is thermally allowed. The stereochemical consequences of the two processes are quite different: the first leads to retention of configuration at the migrating group, and the second leads to inversion of configuration.

Similarly, all groups which can make use of a p-type orbital in the transition state have this alternative route open to them. The same applies to shifts of higher order; suprafacial shifts which are unfavourable for hydrogen are possible for other groups which can use p-type orbitals for bonding in the transition state. The selection rules for these groups are given in table 7.2.

TABLE 7.2. *Selection rules for sigmatropic group migrations of order* $[1, j]$

Number of electrons	Neutral	Polyene Cation	Anion	Thermally allowed migration
2	—	[1, 2]	—	suprafacial, with retention
4	[1, 3]	[1, 4]	[1, 2]	antarafacial, with retention suprafacial, with inversion
6	[1, 5]	[1, 6]	[1, 4]	suprafacial, with retention antarafacial, with inversion
$4n$	$[1, (4n - 1)]$	$[1, 4n]$	$[1, (4n - 2)]$	antarafacial, with retention suprafacial, with inversion
$4n + 2$	$[1, (4n + 1)]$	$[1, (4n + 2)]$	$[1, 4n]$	suprafacial, with retention antarafacial, with inversion

Examples: Neutral species. There are relatively few examples of $[1, j]$ sigmatropic shifts in neutral polyenes involving atoms or groups other than hydrogen. The most likely ones (those for which the transition states are geometrically most accessible) are:

for a thermal [1, 3] shift, suprafacial migration with inversion of configuration, as show in fig. 7.8.

for a thermal [1, 5] shift, suprafacial migration with retention of configuration, just as for hydrogen migration.

The search for systems in which an alkyl group undergoes a [1, 3] shift with inversion has produced some elegant experimental work.[7] One example, due to Roth and Friedrich, is the thermal rearrangement

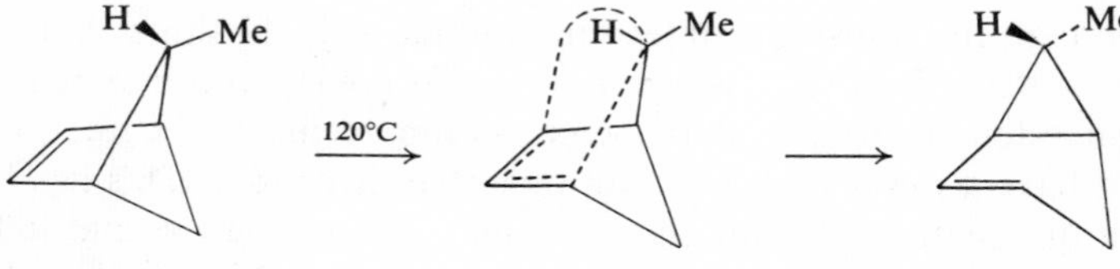

(7.9)

(14)

of the bicyclohexene (14, 7.9). The substituted methylene bridge migrates in such a way that the methyl group is almost exclusively *exo* in the product; if the migration involved retention at the bridging group, the methyl would be *endo*.

It is doubtful whether this mechanism is of any great generality; even in strained systems like (14), steric effects of substituents can prevent the formation of the cyclic transition state necessary for the concerted reaction. An alternative is a stepwise pathway involving radical intermediates. Such a pathway is likely to be especially favourable when the radicals can be stabilised by delocalisation (7.10).

$$\text{R} \quad \rightleftharpoons \quad \text{R} \cdot \quad \rightleftharpoons \quad \text{R} \qquad (7.10)$$

Radical dissociation–recombination mechanisms of this type can be detected directly by the observation of chemically induced dynamic nuclear polarisation (CIDNP) signals in the n.m.r. (see chapter 1). The method has been used to establish the radical mechanism for 1,3-allylic shifts (15) to (16) where the radical intermediate is stabilised by the lone pair on the substituent atom X (7.11).[8]

$$CH_2X \quad \rightleftharpoons \quad \cdot CH_2X \quad \rightleftharpoons \quad CH_2X \qquad (7.11)$$

$$\qquad (15) \qquad\qquad\qquad\qquad\qquad\qquad (16)$$

Thus, both concerted and stepwise mechanisms have been shown to operate in 1,3-allylic rearrangements. In these systems, the validity of the theory of concerted reactions remains intact, but the results emphasise the point that the rules are only *selection* rules; that is, they only predict the preferred stereochemistry of a concerted process and do not exclude the possibility that the reaction may go by a stepwise path of lower energy than any concerted path.

Although the transition states for suprafacial [1, 5] shifts should be relatively strain free, [1, 5] alkyl or aryl shifts are less common than the corresponding hydrogen shifts. One reason seems to be that these groups have a lower migratory aptitude than hydrogen. This has been shown in equation 7.12 for the particular case of the thermal indene rearrangements (17) to (18). In these the tendency of the group X to

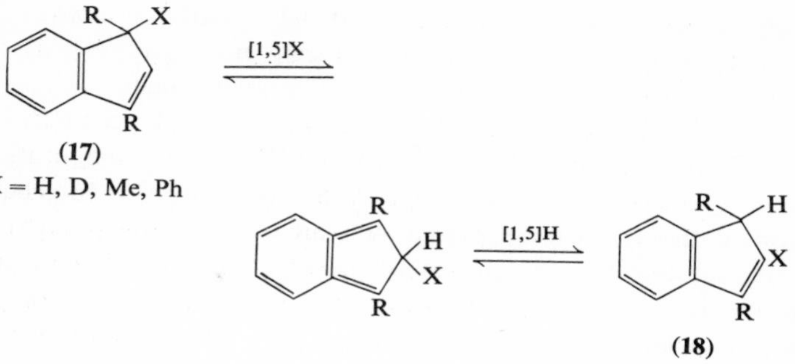

(17)

X = H, D, Me, Ph

(7.12)

(18)

migrate was in the order $H \gg Ph > Me$; the bridging ability of the hydrogen s-orbital in the transition state must be greater than that of the hybrid phenyl or methyl orbitals.[9a]

The stereochemistry of the [1, 5] alkyl shifts is as predicted by the rules. The spirodienes shown in equations 7.13a and b rearrange stereospecifically, with retention of configuration at the migrating group.[9b]

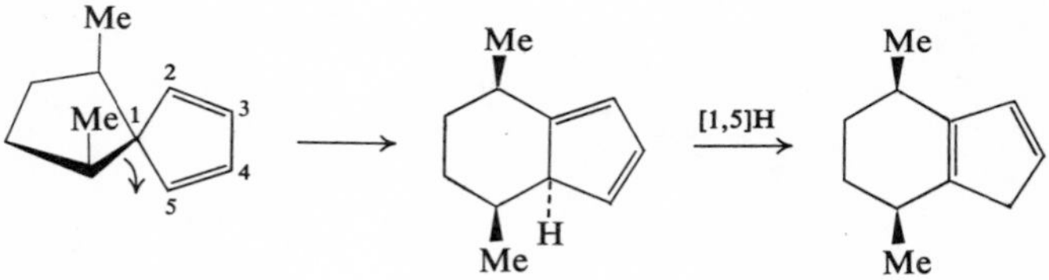

(7.13a)

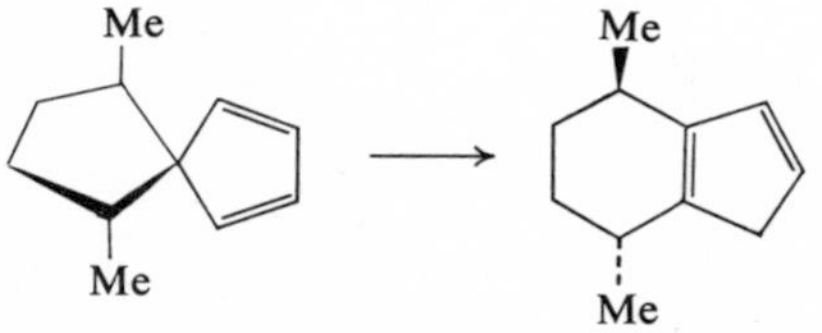

(7.13b)

Charged species. Alkyl or aryl shifts in carbonium ions and other electron deficient species represent a large and important group of molecular rearrangements. In this group are Wagner–Meerwein rearrangements of terpenes, the dienone–phenol rearrangement, the pinacol–pinacolone rearrangement, and expansion and contraction of small rings via carbonium ion intermediates. The electron deficient centre need not be a carbonium ion; the requirement is only that it has a vacant p-orbital. Thus, [1, 2] shifts in singlet carbenes, nitrenes and nitrenium ions are of the same general type (fig. 7.9). The group therefore also includes rearrangement to electron deficient carbon, such as the Wolff rearrangement, and to electron deficient nitrogen, such as the Hofmann, Beckmann, Curtius, and Schmidt rearrangements.

The simple theory predicts that [1, 2] shifts to electron deficient centres should proceed suprafacially and with retention of configuration, through a two-electron Hückel type transition state.[10a]. Molecular orbital calculations have been carried out which support this prediction. It is an oversimplification to represent all these reactions as going through free intermediates of the type shown in fig. 7.9, which then rearrange; the anion associated with the intermediate may play a major part in determining the product distribution, and in some cases the [1, 2] shift and the loss of the anion are probably concerted. The bridged species may even be an intermediate, especially if it involves a migrating aryl group. Despite these variations, the simple qualitative picture is useful. In these reactions the migrating group moves with retention of configuration, as predicted. This has been shown for a carbonium ion rearrangement[10b] and for several other [1, 2] shifts. For example,

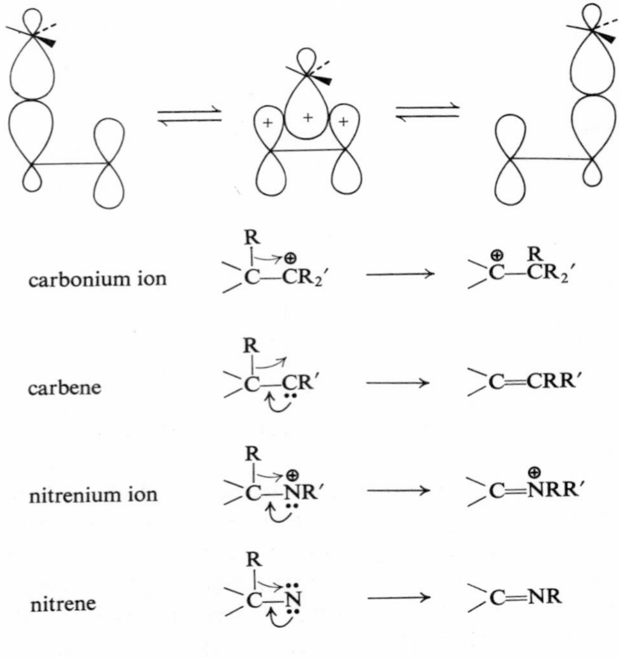

Fig. 7.9. [1, 2] Shifts to electron deficient centres.

Hofmann and Curtius degradations of bridgehead amides proceed normally; neither inversion nor racemisation is sterically possible at a bridgehead carbon.

The theory is also useful in that it predicts that concerted [1, 2] shifts to electron rich centres should be unfavourable. This is reflected in the rarity of 1,2-shifts of hydrogen or alkyl groups in radicals and anions; when such rearrangements are observed in anions, they appear to be stepwise (§ 7.5). Aryl migrations are more common, but in these cases the picture is more complicated because of the different possible orbital arrangements with a bridging aryl group in the transition state.[10a]

[1, 4] Shifts in cationic polyenes are allowed either as antarafacial processes where the configuration of the migrating group is retained, or as suprafacial processes with inversion at the migrating centre. These possibilities are illustrated in fig. 7.10. Migrations of this type occur in systems where the migrating group is constrained to move in a suprafacial manner, and the predicted inversion of configuration has been observed. One example is the rearrangement of the bicyclohexenyl cation (19).[11]

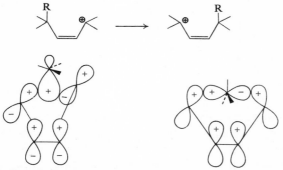

Fig. 7.10. Cationic [1, 4] shift. (Left) Antarafacial migration with retention of configuration; (right) Suprafacial migration with inversion of configuration. Both are Möbius systems.

(19)

The selection rules also predict that suprafacial [1, 6] shifts should be allowed in cationic π systems, with retention of configuration at the migrating centre; but it is difficult to conceive of a case where [1, 2] shifts would not supervene. One example of a rearrangement which can be regarded either as a [1, 2] shift or as a [1, 6] shift is the migration of the methyl group in the labelled benzenonium ion (**20**). The label is 'scrambled' over the benzene ring.[12]

(**20**)

7.4. [3, 3] Sigmatropic changes. The Cope and Claisen rearrangements

Selection rules. [3, 3] Sigmatropic changes are an important group of thermal rearrangements which involve a six-membered cyclic transition state. This transition state (**21**) can be considered as two interacting allyl systems.

(**21**)

Various geometries are possible for the transition state, which can be classified according to whether each of the allyl systems interacts with lobes of the other system on the same side (suprafacially) or on opposite sides (antarafacially). Fig. 7.11 shows these transition states, the signs in this case being put in to indicate maximum overlap of the interacting lobes. On this basis, the transition states are classed as Hückel systems or as Möbius systems. The 'aromatic transition state' approach to the selection rules is again the most useful here.

The selection rules for the [3, 3] shifts simply follow: the Hückel systems are thermally favoured, and the Möbius system is not. Of the three possible types of transition state for thermal rearrangements, the antarafacial,antarafacial one is much less likely to be found than the others, because it involves twisting of the allyl systems. The chair and boat forms of the suprafacial, suprafacial transition state are both relatively strain free. Of the two, the chair form might be expected to be

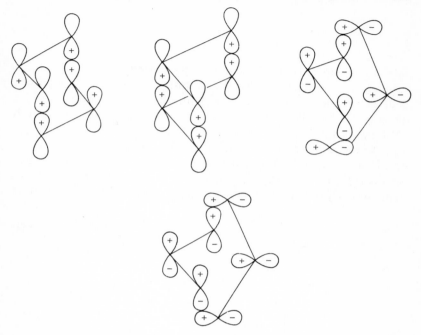

Fig. 7.11. Transition states for [3, 3] shifts. (Top, left to right) Suprafacial, suprafacial (chair); (boat); antarafacial, antarafacial (all Hückel systems). (Below) Suprafacial, antarafacial (Möbius system).

more favoured because the six p-lobes lie in a quasi-planar arrangement. The interaction of the central p-lobes of the two allyl systems in the boat form does appear to have a slight destabilising effect: this has been shown by Woodward and Hoffmann with the aid of correlation diagrams, and by Dewar, by means of molecular orbital calculations.[13] Both the chair and the boat forms are found in practice, as the examples will show.

The Cope rearrangement.[14] 1,5-Dienes isomerise on heating to temperatures up to about 300°C. The reaction is normally reversible and gives an equilibrium mixture of starting material and product. The temperature needed to bring about the reaction depends on the substituents; a conjugating substituent R (acyl, phenyl, etc.) lowers the energy of the transition state and the rearrangement (7.14) goes at 165–185°C.

The reaction shows characteristics typical of a concerted process. It has a large negative entropy of activation, is relatively insensitive to

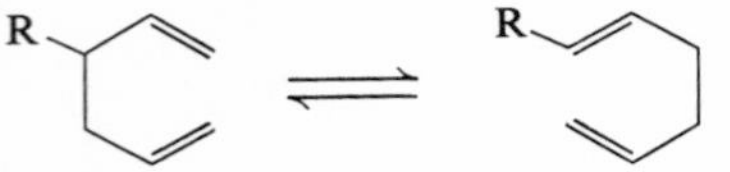

$$(7.14)$$

substituent and solvent effects, and is highly stereoselective. In acyclic 1,5-dienes, the evidence is that the transition state prefers a chair to a boat conformation. Doering and Roth showed this with *meso*-3,4-dimethyl-1,5-hexadiene (22) which rearranged almost exclusively (99.7 per cent) to *cis,trans*-2,4-octadiene at 225°C (7.15 and 7.16).[14] This stereochemistry is consistent only with a chair conformation for the transition state: a boat would give the *cis,cis*- (7.17) or *trans,trans*-octadiene (7.18). Similarly, optical activity is retained in the rearrangement since asymmetry is induced at a new centre. The optically active

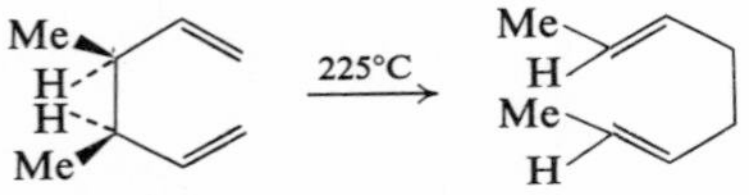

$$(7.15)$$

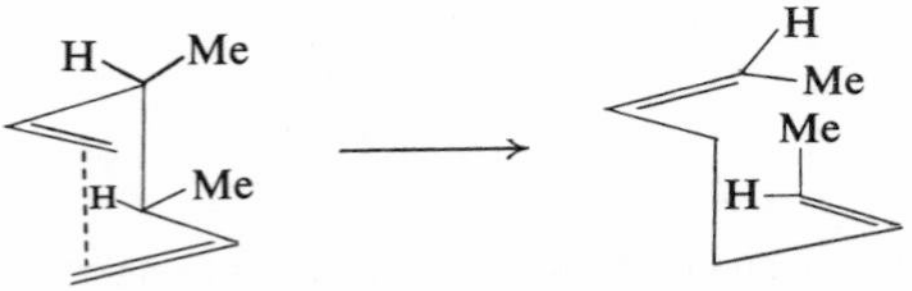

(22)

$$(7.16)$$

cis, trans

but

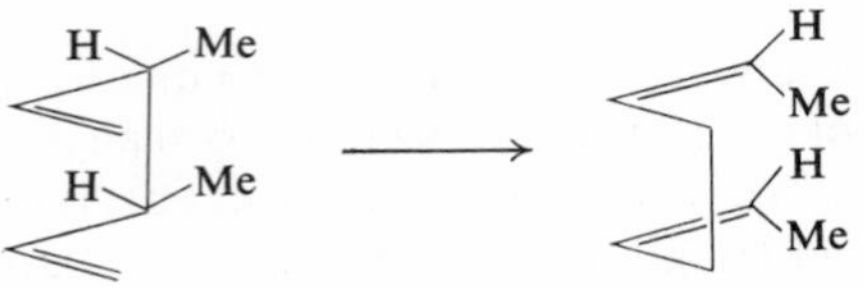

$$(7.17)$$

cis, cis

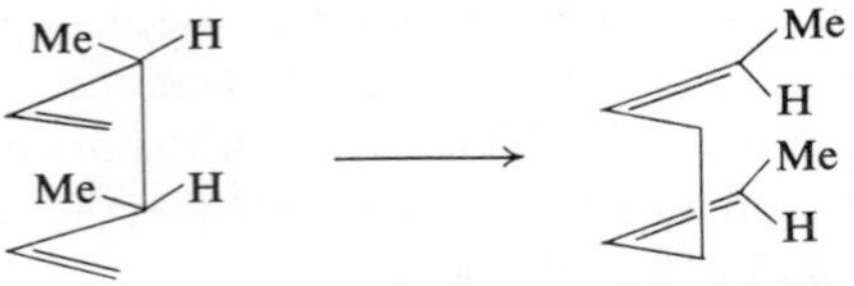

$$(7.18)$$

trans, trans

hexadiene (23) rearranged to an 87:13 mixture of the new hexadienes (24) and (25), both of which were of about 90 per cent optical purity (7.19, 7.20).[15] The stereochemistry and absolute configurations of the products are consistent with the chair conformations shown for the reaction and place an upper limit of about 3 per cent on the contribution of boat conformations. The difference in energy of the two types of transition state in the Cope rearrangement is probably about 6 kcal mol^{-1} (25 kJ mol^{-1}).[13]

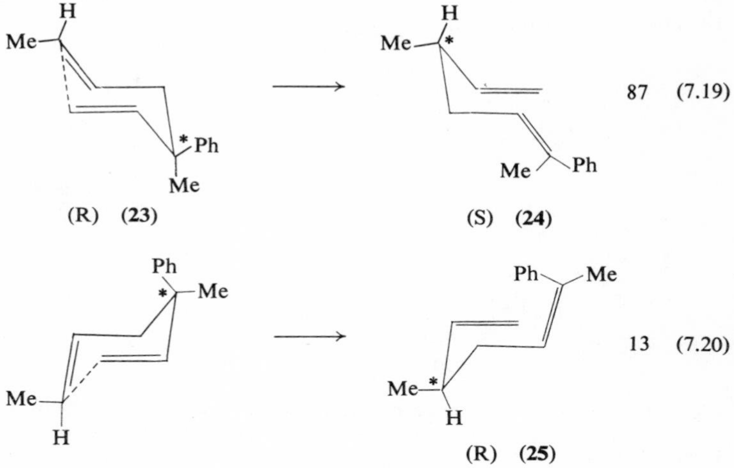

87   (7.19)

(R)   (23)   (S)   (24)

13   (7.20)

(R)   (25)

In some cyclic systems the chair transition state is sterically impossible to attain, and the Cope reaction still goes but by a boat transition state. The Cope rearrangement of *cis*-1,2-divinylcyclopropane (26) and *cis*-1,2-divinylcyclobutane (27) must involve boat transition states, but both go extremely readily because of the relief of strain in the small rings (7.21, 7.22). The rearrangement of divinylcyclopropane (26) can be regarded as an electrocyclic ring closure in which the cyclopropane ring takes the place of a double bond; however, reactions of this sort have normally been classed as Cope rearrangements.[16]

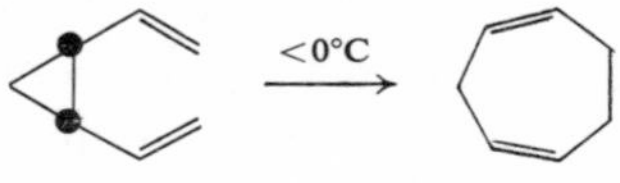

(26)   (7.21)

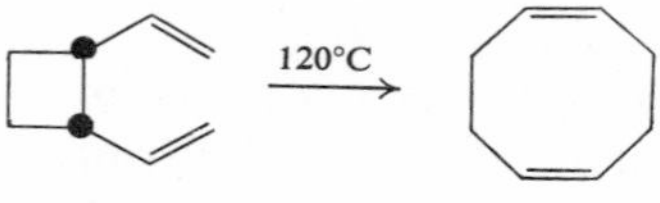

(27)   (7.22)

There are several systems related to (**26**) which undergo Cope rearrangement very readily, and in which the products have the same structure as the starting materials. Compounds which undergo such 'degenerate' rearrangements include the fused cyclopropane (**28**) in the *cisoid* form, and a series of related structures (**29**) in which a bridging group maintains the *cisoid* form (fig. 7.12).

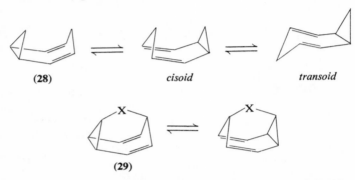

(**28**)   *cisoid*   *transoid*

(**29**)

Fig. 7.12. Degenerate Cope rearrangements for the following: bullvalene (X = —CH=CH—); dihydrobullvalene (X = —CH$_2$CH$_2$—); barbaralane (X = —CH$_2$—); barbaralone (X = >C=O); semibullvalene (no carbon bridge); azabullvalenes (X = —CR=N—).

Dihydrobullvalene, semibullvalene, barbaralane and barbaralone each have two identical valence tautomers. For bullvalene, however, there are not two, but 1,209,600 possible arrangements; the cyclopropane can be at any three adjacent carbons (7.23).

etc.

                                                                         (7.23)

Azabullvalenes (**30**) have potentially the same number of interconvertible forms as bullvalene, but now not all of these are equivalent in energy. The rearrangements have been found to show a strong preference for structures in which the nitrogen is at the junction of a double bond rather than at a bridgehead, so that the number of attainable arrangements is reduced to 28 (7.24).[17]

These rearrangements bear a close relationship to the electrocyclic interconversion of norcardienes and cycloheptatrienes and similar systems, and as in these cases, most of the information about them has

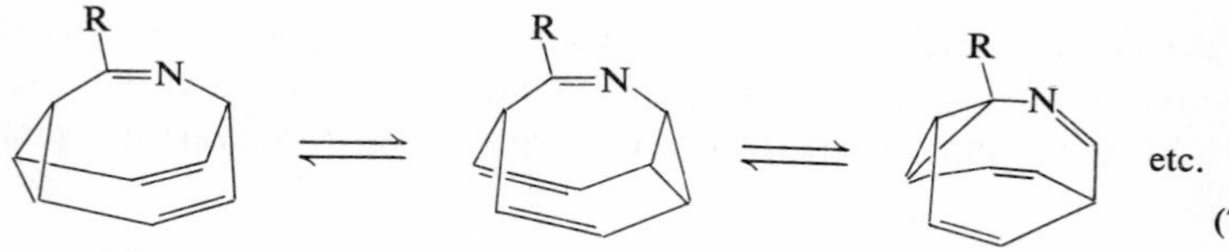

(7.24)

(30)

come from a study of the effect of temperature change on the n.m.r. spectra. The principle of the method has been described earlier (chapter 3). With bullvalene, for example, an n.m.r. spectrum corresponding to the 'frozen' structure is obtained at $-85°C$; there are two sharp signals in the ratio 3:2, the larger (at $\tau = 4.35$) corresponding to the six olefinic hydrogens and the smaller (at $\tau = 7.42$) to the three cyclopropyl and the bridgehead hydrogens. As the temperature is increased these signals gradually become broader and less distinct, and are then replaced by a new maximum which appears as a single sharp peak ($\tau = 5.78$) at $100°C$ and above. Above $100°C$ therefore, the interconversion of the bullvalene tautomers occurs too rapidly for the individual forms to be distinguished, and an average structure, in which all the hydrogens are equivalent, is recorded. Similar averaging processes have been observed in the n.m.r. spectra of the other bridged species **(29)**.

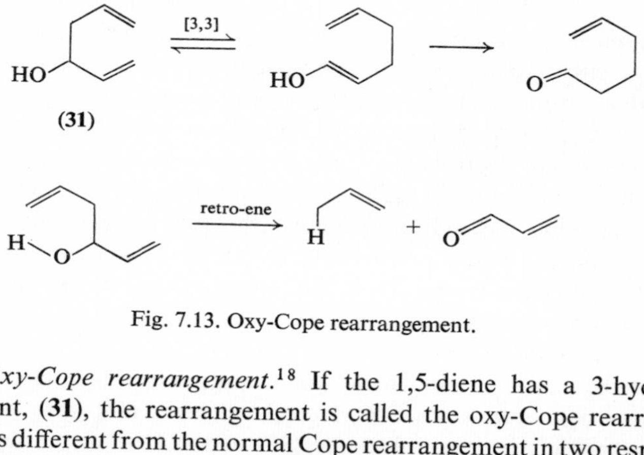

(31)

Fig. 7.13. Oxy-Cope rearrangement.

The Oxy-Cope rearrangement.[18] If the 1,5-diene has a 3-hydroxy substituent, **(31)**, the rearrangement is called the oxy-Cope rearrangement. It is different from the normal Cope rearrangement in two respects: (i) the primary product, an enol, is not isolated since it rapidly tautomerises to the corresponding ketone; and (ii) the [3, 3] shift competes with a retro-ene reaction (§ 7.8), so that the products of both reactions are normally isolated (fig. 7.13).

The Claisen rearrangement.[19] The Claisen rearrangement is basically a thermal [3, 3] sigmatropic rearrangement of an allyl vinyl ether (fig. 7.14). With aliphatic ethers it closely resembles the Cope rearrangement.

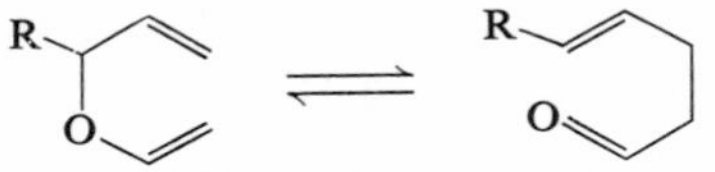

Fig. 7.14

The rearrangement is better known with allyl aryl ethers, however (fig. 7.15). In these systems the reaction is more complex because the product of rearrangement, an *ortho*-dienone, almost invariably reacts further.

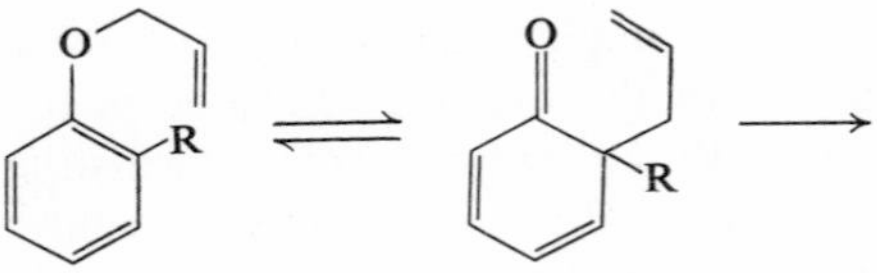

Fig. 7.15

Both in the alkyl and in the aryl series, the reactions show characteristics typical of concerted processes. They are first order and have large negative entropies of activation consistent with cyclic transition states. The allylic group is inverted in the rearrangement, and optical activity is retained if the starting ether is optically active. Polar solvents can increase the rate appreciably – the rearrangement has been observed to be about a hundred times as fast in a polar solvent as in a non-polar solvent – and this may indicate some charge separation in the transition state. Similarly, there is a noticeable, but small, substituent effect on the rate; electron releasing groups in the *para* position increase the rate, and electron withdrawing groups decrease it. The overall rate difference is only about 20 or 30, however.

Like the Cope rearrangement, the Claisen seems to prefer the chair conformation. *trans,trans*-Crotyl propenyl ether (32) gave more than 97 per cent of the *threo*-aldehyde (33) in the Claisen rearrangement, indicating a preference for the chair transition state (7.25). Reaction in the boat conformation leads to the *erythro*-aldehyde (7.26).

Although the Claisen reaction is formally an equilibrium, the disparity in energies of the starting material and product in aliphatic systems is

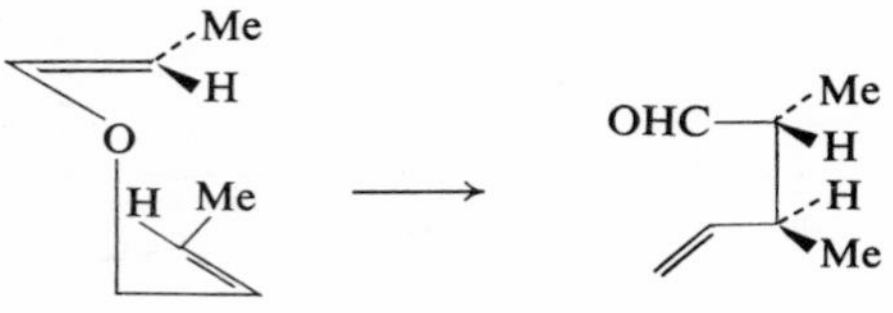

(7.25)

(32)   (33) threo

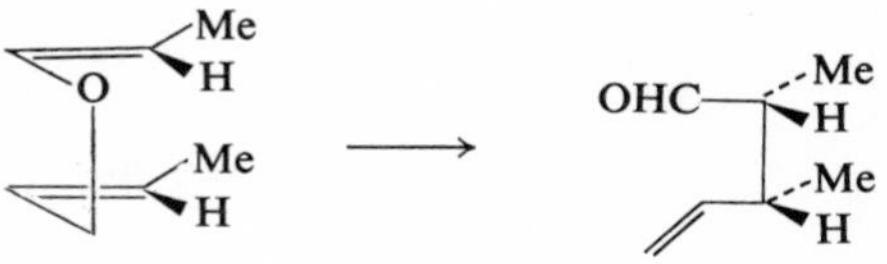

(7.26)

erythro

usually such that the forward reaction goes virtually to completion. This seems to be due mainly to the energy gained in forming the carbonyl group. Even in cases where the reverse reaction (the retro-Claisen rearrangement) would seem to be especially favourable, there is an equilibrium in which the carbonyl tautomer predominates. For example, the equilibrium between 2,5-dihydro-oxepin (34) and its Claisen re-arrangement product, cis-2-vinylcyclopropane carboxaldehyde (35), lies to the right (7.27), even though the reaction involves the formation of the cyclopropane ring; at room temperature the equilibrium ratio of (34) to (35) is 1:19.[20]

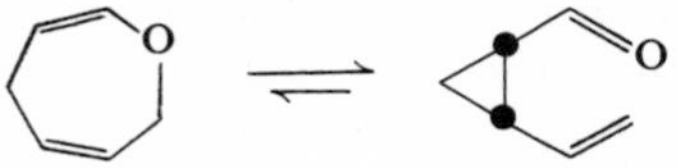

(7.27)

(34)   (35)

With allyl aryl ethers the formal equilibrium is between the allyl ether and the ortho-dienone (fig. 7.16). Here the equilibrium favours the ether, because it is aromatic and the dienone is not. In most cases, how-ever, the dienone is removed as it is formed by other reactions. These are of two main types, illustrated for allyl phenyl ether. One is enolisation to give the o-allylphenol; the other, Cope rearrangement to give the para-dienone which then enolises.

Thus the products normally isolated from aromatic Claisen rearrange-ments are o- and p-allylphenols, the allyl group having undergone one inversion to give the ortho-isomer and two to give the para.

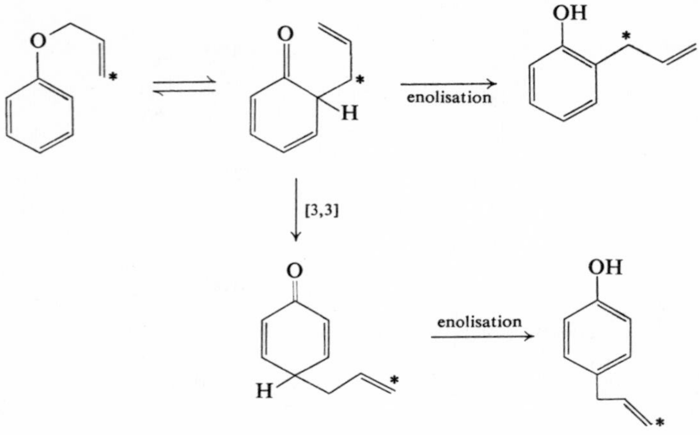

Fig. 7.16. Aromatic Claisen rearrangement.

In unsubstituted phenyl ethers, the enolisation is normally faster than the Cope reaction, so that the product is predominantly or entirely the *ortho*-isomer. On the other hand, when both *ortho*-positions are substituted, enolisation of the *ortho*-dienone is prevented and it undergoes the Cope rearrangement instead. If the *para* position is substituted as well, none of the dienones can enolise and an equilibrium is set up amongst the three dienones and the ether, with the ether predominating.

This general picture may be complicated in individual cases, especially by steric interactions which inhibit the enolisation of the *ortho*-dienone and therefore favour the *para*-isomer. In such cases, enolisation may become the slow step in the formation of the *ortho*-isomer, and so be subject to isotope and solvent effects. For example, the ether (36) rearranges to give both the *ortho*- and *para*-allylphenols. The ratio of products is solvent dependent, less polar solvents favouring the *para*-isomer. The proportion of *para*-isomer is also increased by replacing the aromatic ring protons by deuterium. A possible explanation in this case

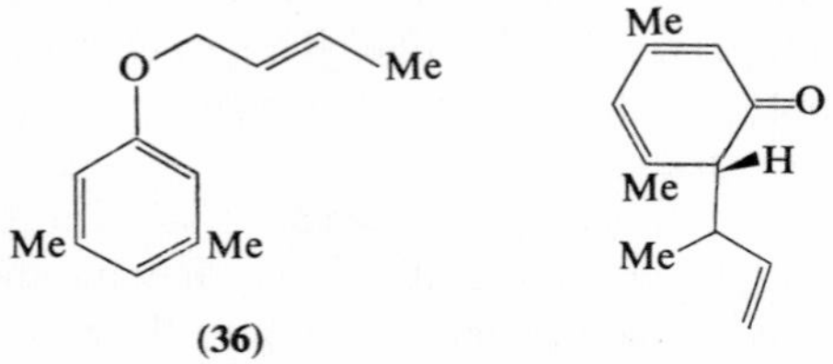

(36)

for the slow enolisation is that in the conformation of the *ortho*-dienone necessary for enolisation, there is a steric interaction between the methyl group on the side chain and the neighbouring ring methyl.

Two other types of reaction that have been observed in aromatic Claisen rearrangements, and are therefore closely linked to them, are not [3, 3] sigmatropic shifts at all. One is the *ortho-ortho*-Claisen rearrangement; the other, the so-called 'abnormal' Claisen rearrangement. These will be described briefly.

ortho-ortho-*Claisen rearrangement.* The Cope rearrangement of the *ortho*- to the *para*-dienone is a common feature of the Claisen. Interconversion of two *ortho*-dienones has also been observed, though it is much less common. An example of the reaction is the formation of the two labelled *ortho*-dienones shown from the labelled mesityl allyl ether (**37**, 7.28). The two *ortho*-dienones are thermally interconvertible (7.29).

(37)   and   (7.28)

(7.29)

If the rearrangement involved a concerted migration through the π system of the ring, it would be a [3, 5] shift, and would be thermally allowed only as a suprafacial, antarafacial process. The geometry of the transition state would be difficult or impossible to attain. It is therefore probably a stepwise reaction. There is some evidence that this is so. The first step is probably an intramolecular Diels–Alder addition of the allyl π bond to the diene, the adduct then undergoing a stepwise fragmentation to the observed product.

Abnormal Claisen rearrangement. Some allyl phenyl ethers with an alkyl substituent on the end carbon of the allyl group, rearrange to give the normal *ortho*-Claisen product plus another isomeric *o*-allylphenol. The second *o*-allylphenol, the 'abnormal' Claisen product, is formed by rearrangement of the normal product; this has been experimentally established (7.30). The abnormal product can be derived from the normal *o*-allylphenol by two [1, 5] homodienyl shifts (§ 7.2). These are illustrated for the system (**38**, 7.31).

e.g.

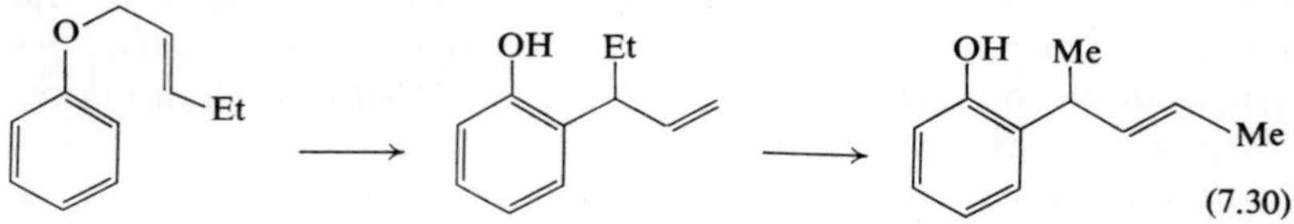

(7.30)

(**38**)

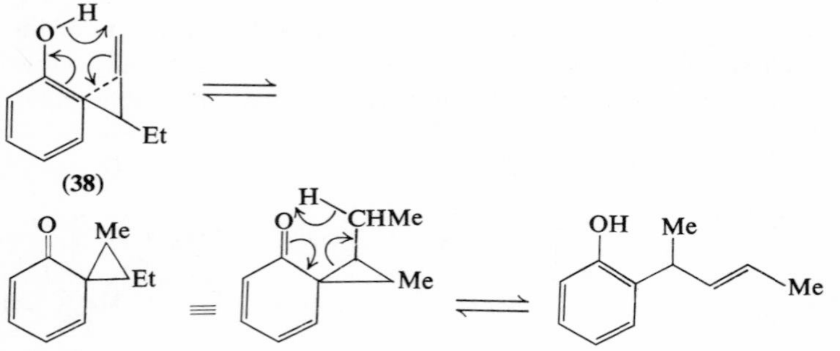

(**38**)

(7.31)

(**39**)

Equilibria of the type (**38**) ⇌ (**39**) may be set up in most Claisen rearrangements, but they cannot lead to the formation of a new product unless there is another alkyl group (here the ethyl group) on the side chain which is able to participate.

A reaction closely related to the abnormal Claisen reaction occurs when phenyl propargyl ether (**40**) is heated (7.32). The product of normal Claisen rearrangement, *o*-allenylphenol, can further rearrange by a [1, 5] hydrogen shift and an electrocyclic ring closure to give a chromene (**41**) which is the observed product.

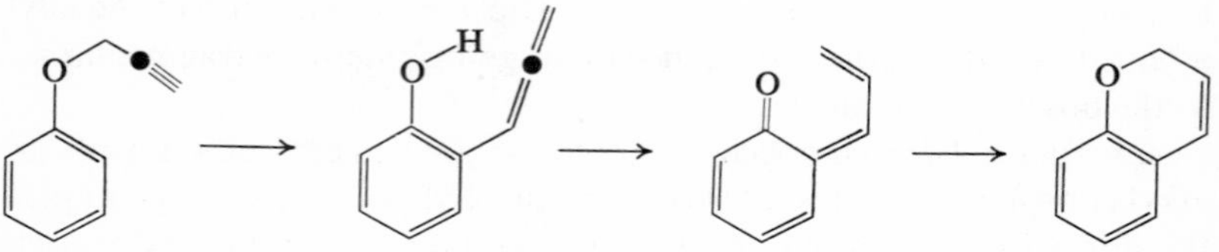

(**40**)

(**41**)

(7.32)

Other [3, 3] *shifts*. Replacement of one or more of the carbon atoms of a 1,5-hexadiene by hetero-atoms does not appear to alter the symmetry requirements for the [3, 3] shift. Thus it is possible that concerted thermal rearrangements can take place in any systems of the general form (42), where one or more of the atoms *a* to *f* are hetero-atoms. The effect of introducing hetero-atoms may be to produce charge separation in the transition state, but still to preserve the overall characteristics of the Cope reaction,[21] as has already been shown with the Claisen rearrangement.

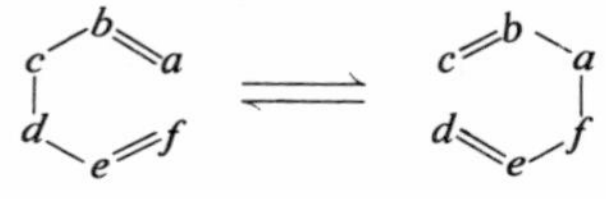

(42)

A few examples where a concerted [3, 3] shift seems the most probable mechanism will serve to illustrate the diversity of systems that can react in this way.

1. The base-catalysed rearrangement of allyl esters to carboxylate anions superficially bears little resemblance to the Claisen rearrangement. However, the transition state is probably of the Claisen type. Equation (7.33) shows an example of the reaction.[22]

$$Me_2CHCOOCH_2CMe{=}CH_2 \xrightarrow{NaH} \qquad \longrightarrow \qquad (7.33)$$

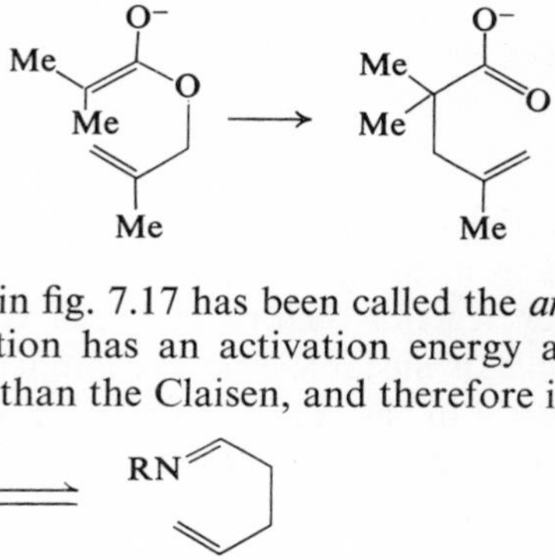

2. The general reaction shown in fig. 7.17 has been called the *amino-Claisen* rearrangement. The reaction has an activation energy about 6 kcal mol^{-1} (25 kJ mol^{-1}) higher than the Claisen, and therefore is not

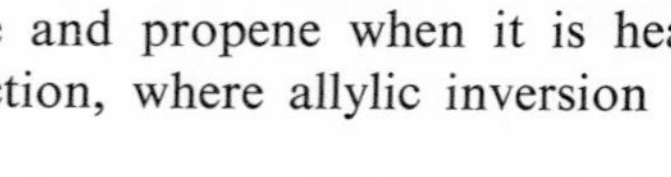

Fig. 7.17

so generally observed, because other reactions may compete at the temperature required for rearrangement. For example, *N*-allylaniline gives mainly aniline and propene when it is heated to 275°C. One example of the reaction, where allylic inversion occurs, is shown in equation 7.34.[23]

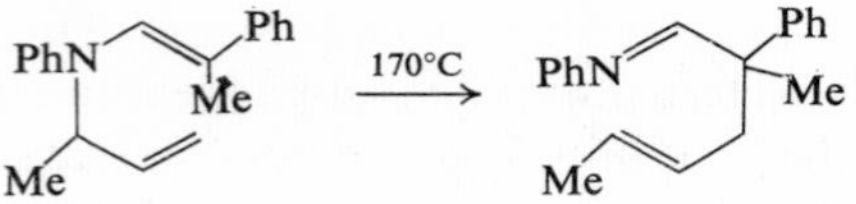

(7.34)

3. The rearrangement shown in fig. 7.18 is the *thio-Claisen* rearrangement.[24] The reaction is much less well known than the Claisen though it is probably just as general and may have an even lower activation

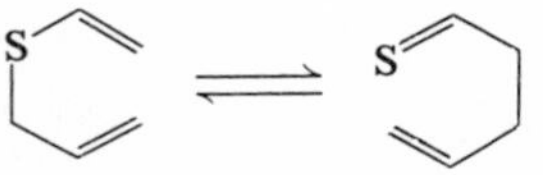

Fig. 7.18

energy. Interpretation of the mechanistic data is complicated by the fact that several other reactions compete with the [3, 3] shift. One important competing reaction is a 1,3-thioallylic rearrangement of the allyl aryl sulphides, for which a mechanism involving a dipolar intermediate has been suggested; the reaction can be intermolecular, giving crossover products with a mixture of allyl aryl sulphides (7.35).

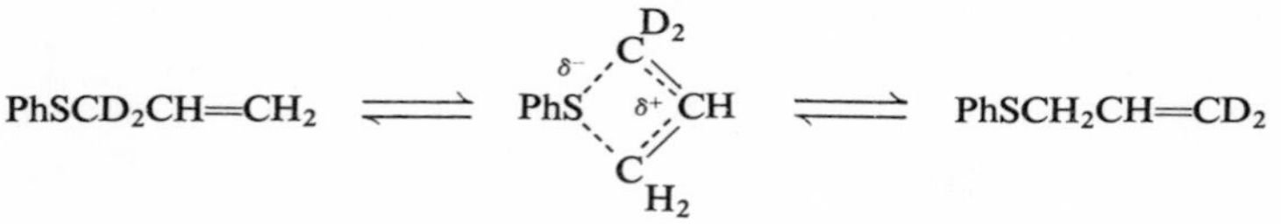

(7.35)

Another complication is that the primary products of the thio-Claisen rearrangement are unstable in the reaction conditions and undergo further reaction before they can be isolated. *o*-Allylthiophenol, for example, is unstable even at room temperature. It can only be detected indirectly in the thio-Claisen rearrangement of allyl phenyl sulphide, and cyclisation products are isolated (7.36).

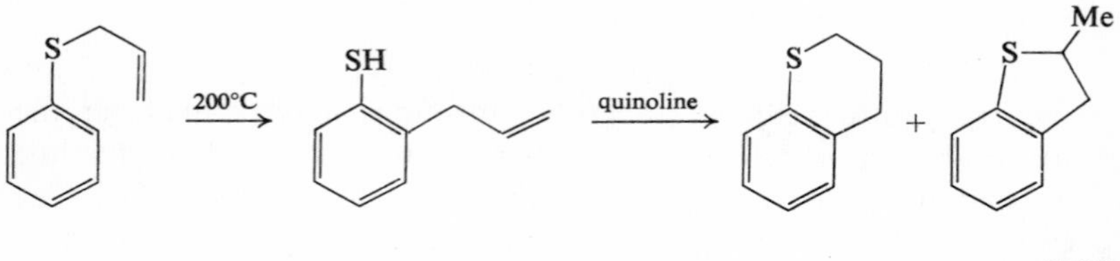

(7.36)

Like the Claisen rearrangement, this rearrangement also takes place with propargyl groups. The sulphide (**43**), for example, rearranges above 80°C to the allene (**44**) which cyclises in pyridine (7.37).

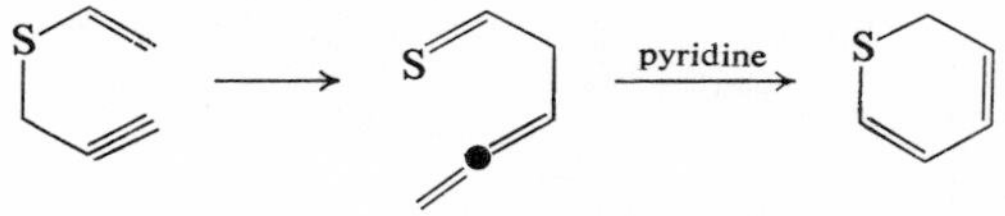

$$(7.37)$$

(**43**)   (**44**)

4. The allylic ester rearrangements (fig. 7.19) are examples of a system in which the transition state for the [3, 3] shift has considerable charge separation. The concerted nature of the process is supported by ^{18}O

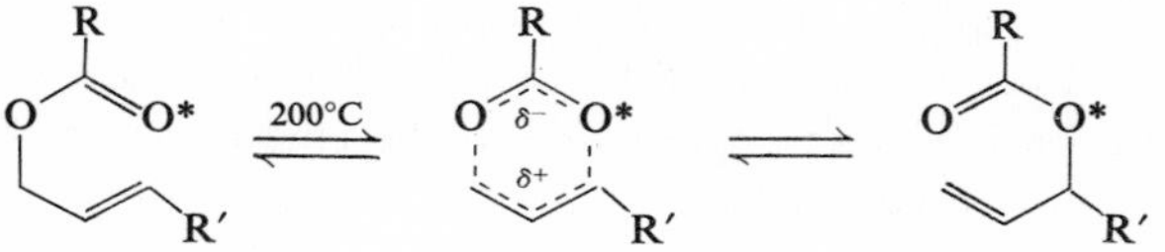

Fig. 7.19

labelling experiments.[25] Similar labelling experiments have been used to establish the [3, 3] shift as a major mechanism for the rearrangement of acetyl peroxide (**45**, 7.38).[26] In both these systems, the concerted mechanism is preferred over the radical dissociation–recombination mechanism (**15** to **16**), which competes so effectively with [1, 3] shifts. Here, the six-membered cyclic transition state must present a lower energy pathway for rearrangement than the radical process.

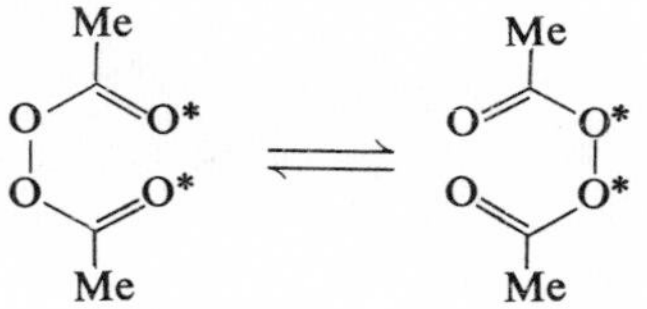

$$(7.38)$$

(**45**)

5. Two rearrangements where a concerted [3, 3] mechanism is suggested by allylic inversion are those of allyl thionocarbonates (**46**, 7.39),[27] and of iminoesters (**47**, 7.40).[28] The latter is an allylic version of the *Chapman* rearrangement. In these systems the equilibrium lies strongly to the right, the driving force being the formation of the carbonyl groups.

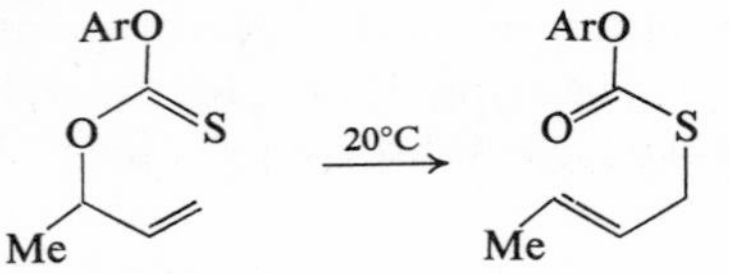

$$ (46) \xrightarrow{20°C} \tag{7.39} $$

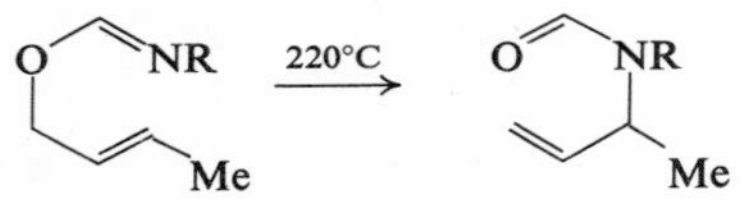

$$ (47) \xrightarrow{220°C} \tag{7.40} $$

6. The Fischer indole synthesis, which is one of the best general methods for preparing indoles, involves the acid catalysed cyclisation of phenylhydrazones.[29] The role of the acid catalyst is probably to assist the tautomerisation of the hydrazone to the enehydrazine (48), but a catalyst is not essential in all cases (7.41). Several phenylhydrazones give indoles simply on heating in a high boiling solvent. The key step, in which the new C—C bond is formed, may be a [3, 3] rearrangement of (48). Positive evidence for the concerted rearrangement is not strong, but an alternative free radical mechanism has been shown to be unlikely.

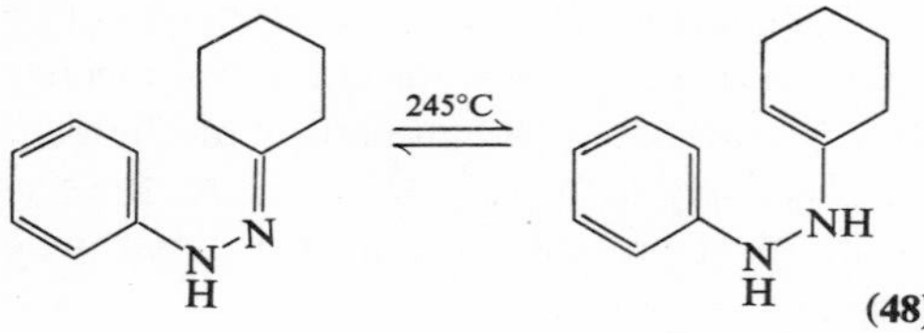

$$ \xrightleftharpoons{245°C} \quad (48) \tag{7.41} $$

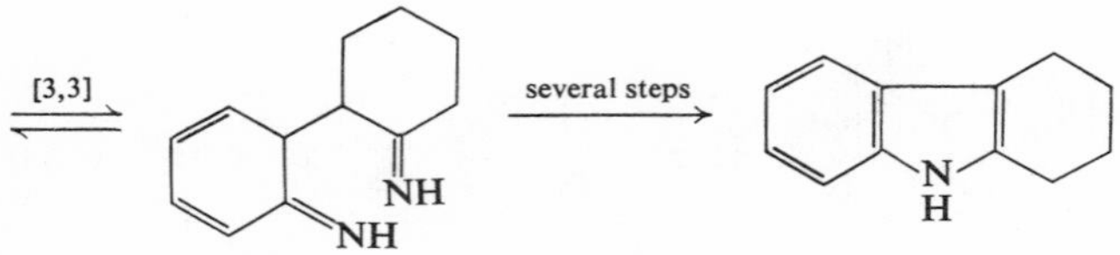

$$ \xrightleftharpoons{[3,3]} \xrightarrow{\text{several steps}} \quad (83\%) $$

Apart from these systems there are several others of the general type (42) which undergo the rearrangement, but for which the mechanism is unknown. The concerted reaction may in fact be quite general for systems which can achieve the cyclic six-membered transition state.

7.5. [2, 3] Sigmatropic changes. Ylide rearrangements. An important group of rearrangements, only recently recognised, involves five-membered cyclic transition states and six electrons.

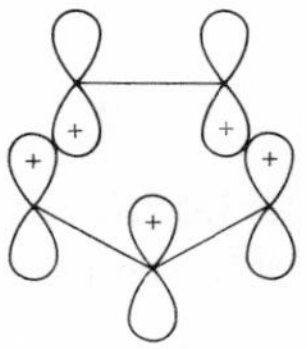

(49)

Fig. 7.20

Consider, as an example, the hypothetical degenerate rearrangement illustrated in fig. 7.20. This is a [2, 3] shift involving six electrons. The transition state (49) for a suprafacial, suprafacial migration is of the Hückel type, and since six electrons participate, the reaction should be thermally allowed via this transition state.

This rearrangement is the anionic equivalent of the Cope rearrangement, and like the [3, 3] shifts, it is not confined to carbon systems. The potential generality of the rearrangement was pointed out by J. E. Baldwin.[30] [2, 3] Shifts involving heterocyclic transition states can be envisaged, represented by the general scheme shown in fig. 7.21 where

Fig. 7.21

one or more of the atoms *a* to *e* are hetero-atoms. Atom *a* need not be an anion; it can be a hetero-atom with a lone pair of electrons available to participate in the reaction.

An example in an all-carbon system is the spontaneous rearrangement of the carbanion (50, 7.42).[30c]

Some examples involving heterocyclic transition states are shown below (equations 7.43–7.49). In some of these, the concerted nature of

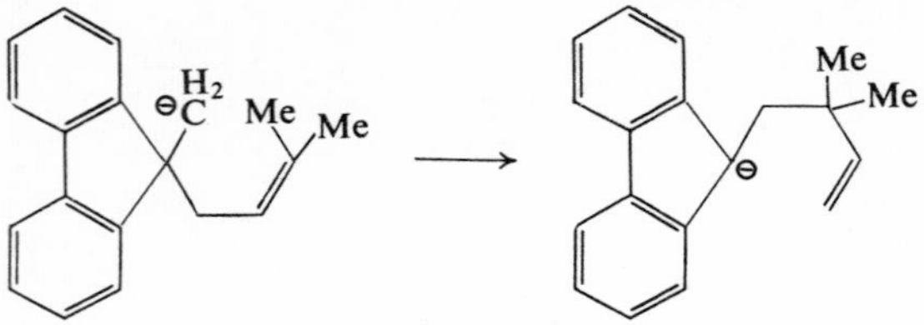

(7.42)

(50)

the rearrangement is supported by the observed inversion of the allyl group. In the others, the [2, 3] shift seems the most likely mechanism for rearrangement, although alternatives may not have been ruled out.

1. Allyl benzyl ether anions (the *Wittig* rearrangement):[30d,31]

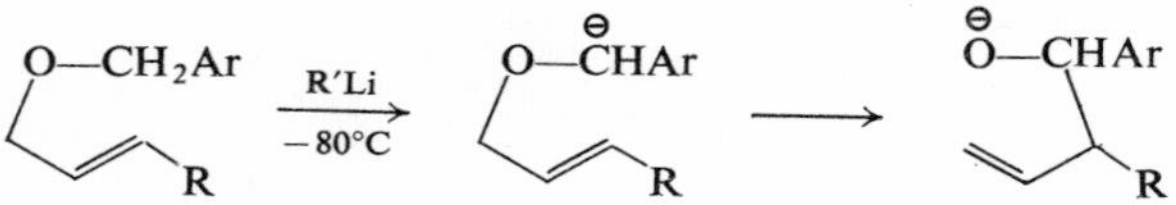

(7.43)

2. Quaternary ammonium ylides (the *Sommelet–Hauser* rearrangement):[31]

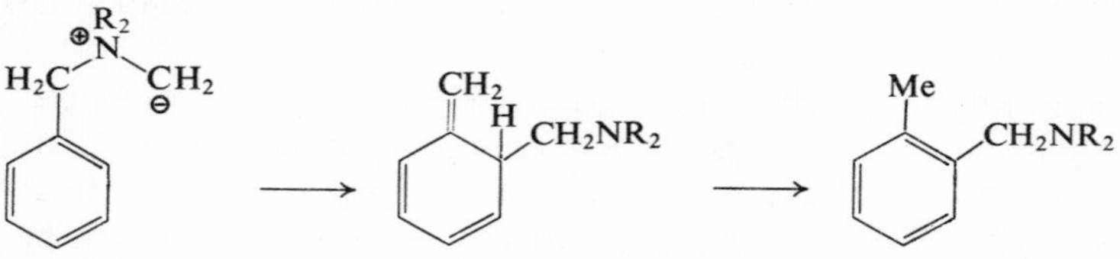

(7.44)

3. Sulphonium ylides:[30a]

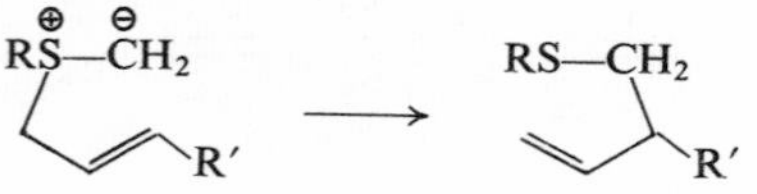

(7.45)

4. Allylic quaternary ammonium oxides (the *Meisenheimer* rearrangement):[32]

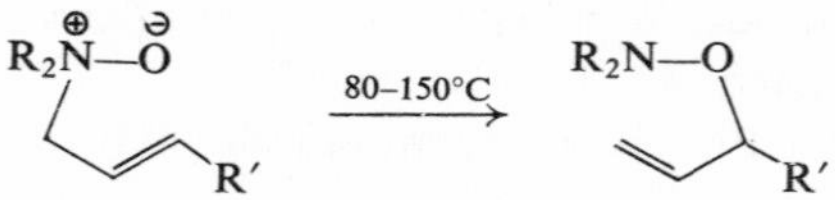

(7.46)

5. Allylic amido-ammonium salts:[33]

$$\overset{\oplus}{R_2N}-\overset{\ominus}{N}COAr \xrightarrow{\quad 120-140°C \quad} R_2N-NCOAr \tag{7.47}$$

6. Allylic phosphinates:[34]

$$O-\overset{..}{P}Ph_2 \longrightarrow \overset{\ominus}{O}-\overset{\oplus}{P}Ph_2 \begin{matrix} D \\ D \end{matrix} \tag{7.48}$$

7. Allylic sulphenates:[35]

$$O-\overset{..}{S}Ar \longrightarrow \overset{\ominus}{O}-\overset{\oplus}{S}Ar \tag{7.49}$$

Some of these rearrangements have been known for many years, while others are of recent origin. The possible variations in the system are enormous, and have by no means been completely explored.

The theory thus accounts very well for rearrangements of this type in which an allyl group migrates with inversion. Some of these rearrangements also occur, however, if the allyl group is replaced by a simple alkyl group. For example, the Wittig rearrangement occurs not only with allyl benzyl ethers, but also with alkyl and aryl benzyl ethers.†

$$\overset{\ominus}{Ar}CHOR \rightarrow ArCHRO^{\ominus} \rightarrow ArCHROH$$

Such rearrangements are 1,2-shifts to an electron rich centre. They can be represented by the general scheme

$$\begin{matrix} \overset{\ominus}{a}-b \\ | \\ c \end{matrix} \longrightarrow \begin{matrix} a-b^{\ominus} \\ | \\ c \end{matrix}$$

Ylide and amine oxide rearrangements of this type also occur; in ammonium or sulphonium ylides, the reaction is known as the *Stevens* rearrangement[31] and with amine oxides, it is the *Meisenheimer* rearrangement.†

$$\begin{matrix} \overset{\ominus}{a}-b^{\oplus} \\ | \\ c \end{matrix} \longrightarrow \begin{matrix} a-b \\ | \\ c \end{matrix}$$

Meisenheimer: $a = O$, $b = NR_2$ Stevens: $a = CR_2$, $b = NR_2$ or SR

Even when the migrating group is an allyl group, a minor product of the reaction is often found in which allylic inversion has not taken place, and which therefore cannot be formed by a concerted [2, 3] shift. These competing routes are illustrated for the Wittig rearrangement of benzyl dimethylallyl ether in fig. 7.22.[36]†

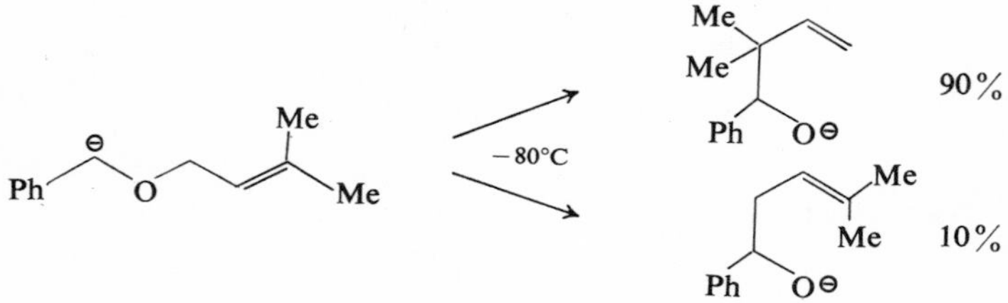

Fig. 7.22. Competing rearrangements of benzyl dimethylallyl ether.

If the principle of orbital symmetry conservation is correct, then these minor reactions must be stepwise processes. One possible stepwise mechanism is a radical dissociation–recombination process, of the type proposed for 1,3-shifts and discussed in § 7.2; it is shown in fig. 7.23 for the Stevens rearrangement.

Fig. 7.23. Radical dissociation–recombination mechanism for Stevens rearrangement.

As for the 1,3-shifts, the best diagnostic test for such a mechanism is the observation of dynamic nuclear polarisation (CIDNP) in the n.m.r. spectrum. This has been observed in Stevens rearrangements. The rearrangement of the ylide (51) has been followed by n.m.r., and a CIDNP signal appears during the reaction at the same place as the quartet of the methine proton in the product (fig. 7.24).[37]

Signal *a* is the quartet of the methine proton of an independently synthesised sample of *N*-methyl-*N*-(α-phenethyl)aniline. Signal *b* is the enhanced signal produced about 30 seconds after generation of the ylide (51), and *c* is the signal 10 minutes later, showing the normal quartet of the proton in the product, but at much lower intensity than that of *b*.

---

† It is unfortunate that for historical reasons, both the concerted (six electron) and stepwise (four electron) rearrangements bear the same names.

$$\overset{\oplus}{Ph\overset{\ominus}{N}Me_2\overset{\ominus}{C}HPh} \rightarrow Ph\overset{\cdot}{N}Me\overset{\cdot}{C}HPh \rightarrow PhNMeCHMePh$$

(51)   Me·

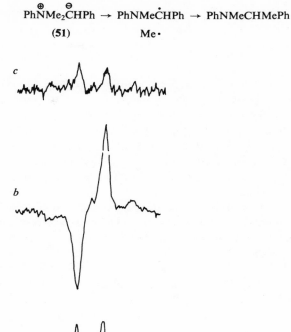

Fig. 7.24. CIDNP signal in Stevens rearrangement. Reprinted from *J. American Chemical Society*, **91**, 1237 (1969).

Signal *b* thus shows two features which characterise it as a CIDNP signal: emission as well as absorption, and enhanced intensity.

Similarly, radical intermediates have been detected by CIDNP in a simple Meisenheimer rearrangement, that of *N,N*-dimethylbenzylamine oxide (fig. 7.25).[38]

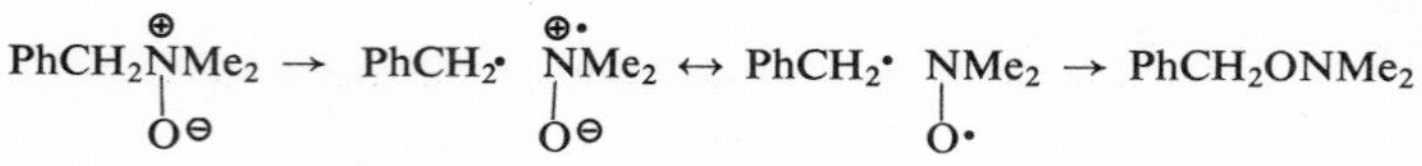

$$\underset{\underset{O\ominus}{\overset{\oplus}{PhCH_2NMe_2}}}{} \rightarrow \underset{\underset{O\ominus}{PhCH_2 \cdot}}{\overset{\oplus \cdot}{NMe_2}} \leftrightarrow \underset{\underset{O \cdot}{PhCH_2 \cdot}}{NMe_2} \rightarrow PhCH_2ONMe_2$$

Fig. 7.25. Radical pairs in Meisenheimer rearrangement.

Though not all the 1,2-rearrangements observed in simple Stevens, Meisenheimer and Wittig rearrangements need involve radical pairs as intermediates, the mechanism does appear to be a common one. When the substrate has an allylic group, the concerted [2, 3] shift competes favourably with the radical process. The concerted process usually has the lower activation energy; this is shown by the fact that at lower temperature, the proportion of the product which is formed by the concerted mechanism increases.

The general pattern of thermal rearrangements therefore seems to be as follows: where the selection rules predict a facile concerted sigmatropic rearrangement, this is the mechanism of rearrangement which is mainly, but not exclusively, followed. Where the transition state for a concerted sigmatropic shift is difficult to attain, stepwise processes increasingly compete; and where rearrangements are observed which are 'forbidden' as concerted processes, there is evidence that they do go by stepwise mechanisms.

7.6. Photochemical rearrangements. Selection rules for photochemical sigmatropic shifts can be devised, starting from the assumption that for the first excited states, $4n$ electron systems of the Hückel type and $(4n + 2)$ electron systems of the Möbius type are the stable ones – the reverse of the ground state situation. Thus, sigmatropic shifts which are unfavourable in the ground state should be favourable in the first excited state, and vice versa.[39]

A suprafacial [1, 3] alkyl shift with retention of configuration provides an example. The transition state contains four electrons and is of the Hückel type (see fig. 7.8), so the reaction is unfavourable in the ground state. In the first excited state the four-electron Hückel transition state is favoured and the reaction should be allowed. Many photochemical reactions do give the products expected from a [1, 3] sigmatropic shift; for example, cyclic unsaturated ketones of the general type (**52**) are found to rearrange photochemically to the four-membered ring isomers (**53**).

This system illustrates a problem with the simple orbital symmetry rationalisation. The carbonyl group is necessary for the reaction to take place, and the initial photochemical excitation probably involves this group. The group could thus simply provide a means of absorbing energy for a concerted, excited state rearrangement not directly involving the carbonyl function, or the reaction could be a stepwise one. Like other photochemical reactions, the mechanism may be more complex than the simple theory suggests.

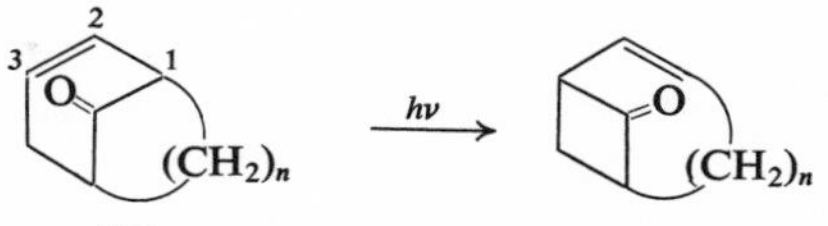

$$\text{(52)} \quad \xrightarrow{h\nu} \quad \text{(53)} \tag{7.50}$$

Metal catalysis might also be a method of bringing about [1,3] shifts; indeed, transition metal carbonyls have catalysed formal [1, 3] hydrogen shifts in allylic systems, but the available evidence points to a stepwise mechanism, with metal hydride intermediates for these reactions.[40] It has not been conclusively shown that metal catalysts can reverse the symmetry rules for thermally forbidden rearrangements, and make them symmetry-allowed processes.

7.7. The ene reaction. The addition of an olefin containing an allylic hydrogen atom to a π bond, in the manner illustrated in fig. 7.26 for propene and ethylene, is the *ene reaction*.[41]

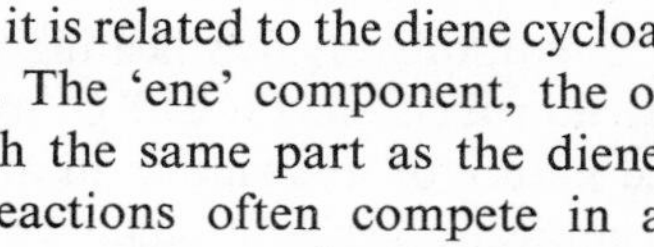

Fig. 7.26. The ene reaction.

As its name implies it is related to the diene cycloaddition reaction, the Diels–Alder reaction. The 'ene' component, the olefin with the allylic hydrogen, takes much the same part as the diene in the Diels–Alder reaction. The two reactions often compete in a particular system. Mechanistically, however, the reaction is much more closely related to [1,5] sigmatropic shifts. It is therefore included in this chapter even though it does not strictly fit the definition of a sigmatropic rearrangement.

Theory. The most likely geometry of approach for the components of an ene reaction is as shown, (**54**). This is very similar to the approach in the Diels–Alder reaction, and will lead to a 'boat like' transition state.

The extra flexibility of the ene system compared to the diene may allow the hydrogen atom to swing down towards the π bond in the transition state, so that the developing p-orbital on the carbon of the C—H bond is parallel to those of the adjacent double bond, as in (55). In any case, the result of this approach geometry will be a *suprafacial* (*cis*) addition to the termini of the π bond.

(54)

(55)

Unlike the Diels–Alder reaction, but like many sigmatropic reactions, the ene reaction does not have a symmetrical transition state, and correlation diagrams are not readily constructed. However, we can deduce that it is thermally allowed as a concerted reaction by analogy with closely related systems: its transition state is 'aromatic' in the sense that it is a Hückel system and involves a suprafacial interaction of six electrons (four from the π bonds and two from the σ bond). In the general terminology of Woodward and Hoffmann it can be regarded as a $(_\sigma2_s + _\pi2_s + _\pi2_s)$ reaction.

Examples of the reaction. The reactivity of the components of the ene reaction often parallels their reactivity in the Diels–Alder reaction. Like the Diels–Alder, the reaction goes best with electron rich ene components and electron deficient enophiles. Maleic anhydride is a good enophile, for example. Typical conditions involve heating it with the ene component in refluxing trichlorobenzene (at 210°C) for 12–24 hours. Evidence for the concerted nature of its reaction in at least one instance is the formation of an optically active adduct with optically active 3-phenylbut-1-ene (7.51). This is an example of asymmetric induction.

(7.51)

Acetylenes with electron withdrawing substituents, such as acetylene-dicarboxylic esters and propiolic esters, are even better enophiles. The reaction intermediate benzyne, which can be regarded as a highly strained acetylene, is a powerful enophile, and with several acyclic dienes it gives products both of the ene and of the Diels–Alder reactions. The ene competes favourably with the Diels–Alder reaction with acyclic dienes because it does not require a *cisoid* conformation of the diene.

If the transition state for a concerted ene reaction is difficult to attain for steric reasons, a stepwise addition may occur instead. In most cases, however, a concerted mechanism is indicated. This has been shown in particular, for ene reactions with diethyl azodicarboxylate (**56**), which is a very good enophile and reacts with ene systems at temperatures below 100°C. In the ene reaction with 1-phenyl-3-(*p*-tolyl)propene (**57**), for example (7.52), there is only a fourfold increase in rate over a range of solvents from cyclohexane to nitrobenzene.

The ene reaction of diethyl azodicarboxylate with 1,4-dihydronaph-thalene at 60–80°C shows a deuterium isotope effect, k_H/k_D, of 2.8 to 4.1, suggesting some C—H bond breaking in the transition state, and again consistent with a concerted mechanism.

The reaction of singlet oxygen (produced by dye-sensitised irradiation of oxygen) with olefins appears to be an example of the ene reaction; the products are those expected from an ene reaction, allylic hydroperoxides (7.53). However, the products are not formed by an ene reaction at all,

$$\text{e.g.} \quad \text{Me}_2\text{C}{=}\text{CMe}_2 + \text{O}_2\,(^1\Delta\text{g}) \rightarrow \text{H}_2\text{C}{=}\text{CMeC(OOH)Me}_2 \qquad (7.53)$$

at least in the system shown (7.54).[42] In the presence of an added nucleo-phile (azide ions), the reaction is almost completely diverted to give the azido-peroxide (**59**). This shows that the peroxide formation cannot be concerted and that there must be an intermediate, susceptible to nucleo-philic attack. A possible structure for the intermediate is the zwitterion (**58**).

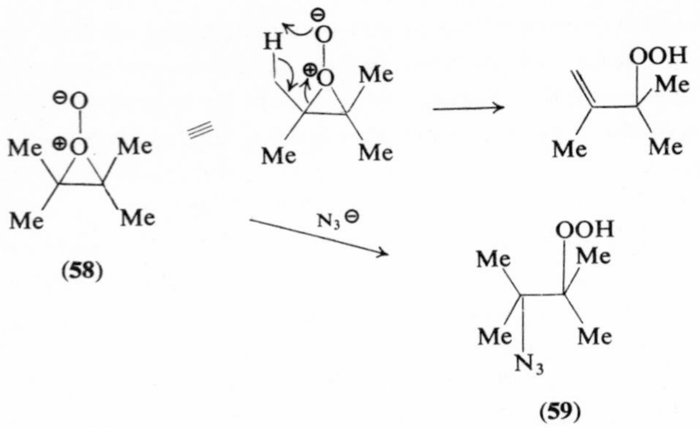

(7.54)

(58)

(59)

Thus, although concerted ene reactions are the rule, in line with the predictions of the theories of concerted reactions, there are alternative stepwise mechanisms of lower activation energy available in a few systems.

7.8. The retro-ene reaction. This fragmentation essentially involves transfer of hydrogen through a six-membered cyclic transition state.[41,43] It bears a close resemblance to the [1, 5] sigmatropic shift, the main difference being that in the retro-ene reaction, a σ bond breaks instead of a π bond. The transition state is 'aromatic' in that it involves six electrons (see the discussion of the mechanism of the ene reaction, § 7.7).

The reaction is not well known in all-carbon systems except where they form part of medium-sized rings. Cyclic olefins with rings of eight to eleven atoms can be converted into terminal dienes by heating to about 500°C, the products being removed as they are formed to prevent further reactions (fig. 7.27). More commonly, one or more of the atoms in the

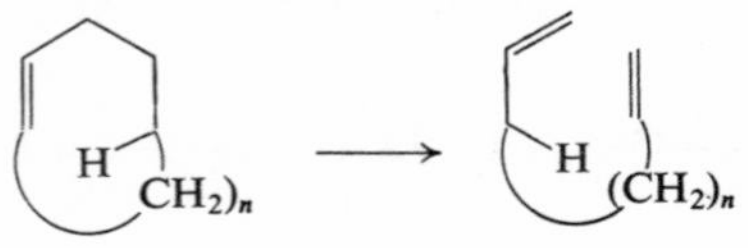

$(n = 3 \text{ to } 6)$

Fig. 7.27

transition state is a hetero-atom, especially oxygen. This group includes some of the important olefin forming reactions and decarboxylations.

Several of these reactions which seem most likely to involve concerted fragmentation via a cyclic transition state will be described briefly.[44]

1. *Pyrolysis of esters.* When esters are heated to 300–500°C, they decompose to an olefin and a carboxylic acid (7.55). The ester pyrolysis

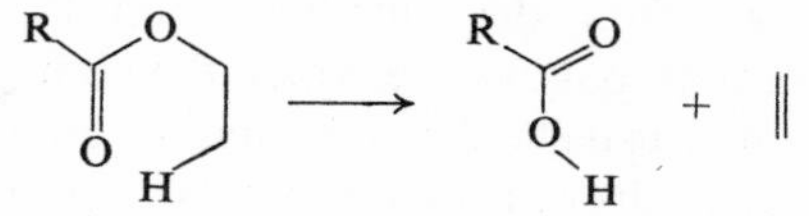

$$R \overset{O}{\underset{O\,H}{\diagup}} \longrightarrow R \overset{O}{\underset{O\,H}{\diagup}} + \parallel \tag{7.55}$$

is stereospecific in the *cis* sense, unimolecular, shows a deuterium isotope effect and generally, a negative entropy of activation. These characteristics are all to be expected for a concerted cyclic elimination. The *cis* stereospecificity has been demonstrated in several systems. An example (fig. 7.28) is the pyrolysis of *threo*- and *erythro*-3-deuterio-2-butyl acetates, (60) and (61). Both give *cis*- and *trans*-2-butenes. From the

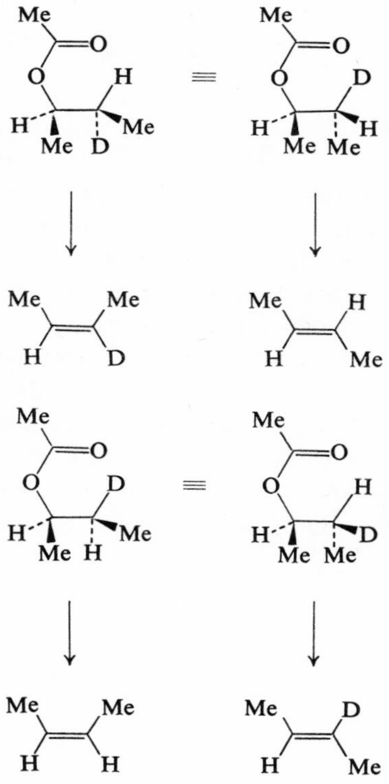

Fig. 7.28

threo isomer (**60**), the *cis*-butene contains deuterium and the *trans*-butene does not; from the *erythro*-isomer (**61**) the *cis*-butene is undeuterated and the *trans*-butene is deuterated, as expected for a *syn* elimination.[45]

Some 1-butene is also formed in each case, by alternative elimination of a primary hydrogen from the α-methyl group. In cases such as this where there is more than one way in which the elimination can go, the ratio of olefins formed usually depends mainly on a statistical effect; that is, the number of hydrogens of a particular type available for elimination. Occasionally, other factors, such as the stabilities of the olefins being formed, become dominant. In cyclic systems, conformational effects are important: if the ester group occupies an axial position in a cyclic system, for example, then the hydrogen on the adjacent carbon must be equatorial for elimination to take place (fig. 7.29). If the leaving

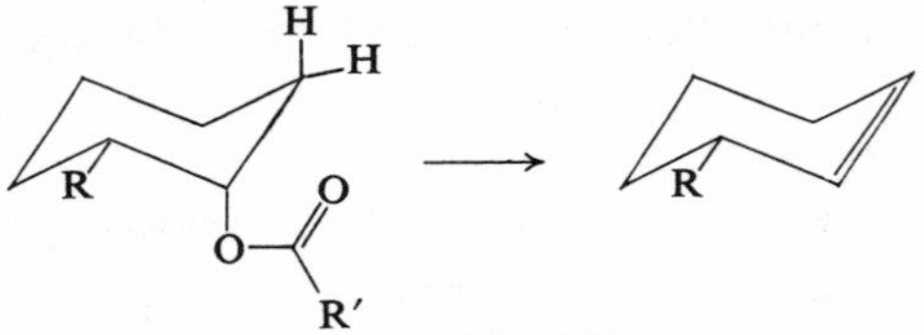

Fig. 7.29

group is equatorial, then both *cis* and *trans* hydrogens on the adjacent carbon are sterically accessible for a cyclic transition state. However, a concerted elimination involving the *trans* hydrogen will give a transition state in which the developing double bond is *transoid*. Again, therefore, *cis* elimination is expected to be favoured energetically.

2. *Pyrolysis of xanthates (Chugaev reaction)*.[44] Xanthate esters are prepared from alcohols by reaction with carbon disulphide and alkali, followed by methylation with iodomethane. Their thermal decomposition follows a pattern similar to that of carboxylate esters but goes at much lower temperatures (100–250°C), probably because of the energy gained in the reorganisation $-O-C{=}S \rightarrow O{=}C-S-$ (7.56).

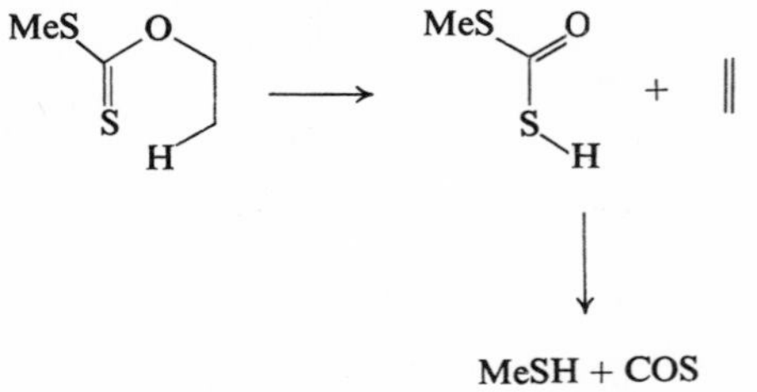

$$(7.56)$$

MeSH + COS

This reaction probably involves a concerted fragmentation via a cyclic transition state, but the evidence is less definite. Usually the elimination is stereospecifically *syn*, but *anti*- eliminations have occasionally been observed, especially when there is an electron withdrawing group at the β-carbon. The transition state may be highly polarised, so that suitable substituents can change the mechanism of fragmentation to a stepwise one.

3. *Decarboxylation of β-keto-acids.*[46] It is well known that β-keto-acids are thermally unstable, and lose carbon dioxide when heated (7.57). Since these acids are intramolecularly hydrogen bonded even in the solid state, a cyclic mechanism seems most likely for the fragmentation.

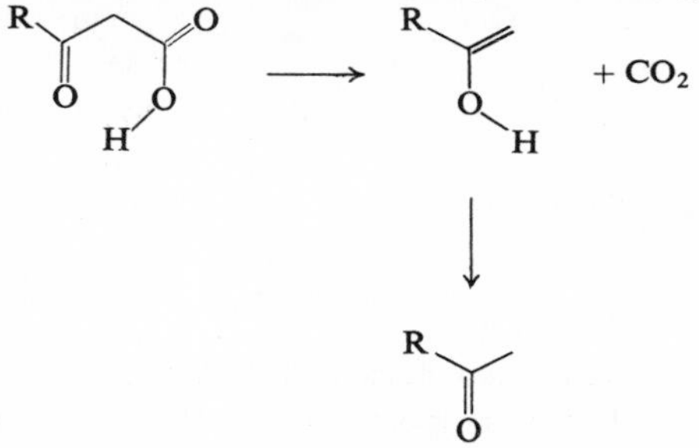

(7.57)

The fact that different solvents have little effect on the rate argues against a dipolar intermediate in the reaction. However, the transition state probably does have some polar character; both deuterium isotope studies and measurements of volumes of activation suggest that there is some transfer of the acid proton to the ketone oxygen as the transition state is approached (fig. 7.30).

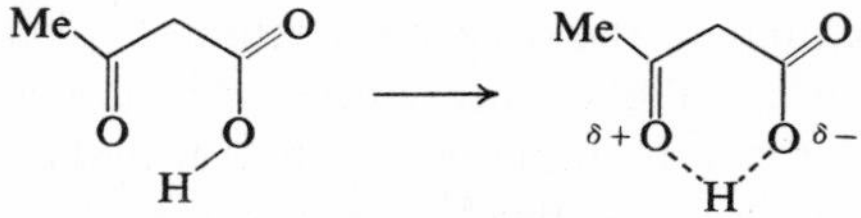

Fig. 7.30

4. *Pyrolysis of β-hydroxy-ketones.*[47a] Like β-keto-acids, a six-membered cyclic hydrogen bonded structure can be drawn for β-hydroxy-ketones. When they are heated to 200–250°C, a thermal retrograde aldol reaction takes place. The vapour phase pyrolysis of diacetone alcohol gives acetone, for example (7.58); the reaction has a negative entropy of activation and an activation energy of 31 kcal mol^{-1} (130 kJ mol^{-1}). A cyclic transition state, probably with some dipolar character, is likely. β-Hydroxy-esters can react in a similar way.[47b]

$$\text{(structure)} \longrightarrow 2 \text{ MeC}=\text{O} \qquad (7.58)$$

5. *Decarboxylation of* β,γ-*unsaturated acids.*[48] β,γ-Unsaturated carboxylic acids lose carbon dioxide when heated above about 300°C (7.59). The reaction is first order, and has a negative entropy of activation. A concerted fragmentation is the most probable mechanism. The product is an olefin in which the double bond has migrated through one carbon – typical of the ene and retro-ene reactions. Cyclohexylacetic acid, for example, gives methylenecyclohexane (7.60).

$$\text{(structure)} \longrightarrow \text{(structure)} + CO_2 \qquad (7.59)$$

$$\text{(structure)} \longrightarrow \text{(structure)} + CO_2 \qquad (7.60)$$

α,β-Unsaturated acids also decarboxylate when heated, but more slowly. They probably first rearrange (by a [1, 5] sigmatropic shift) to the β,γ-unsaturated isomers, which then decarboxylate (7.61). The

$$\text{e.g.} \quad \text{(structure)} \xrightarrow{[1,5]} \text{(structure)} \longrightarrow Me_2C=CH_2 + CO_2 \qquad (7.61)$$

reaction is very similar to the decarboxylation of β-keto-esters but the reaction temperature is higher, presumably because of the absence of intramolecular hydrogen bonding in the unsaturated acids.

6. *Pyrolysis of* β-*hydroxy-olefins.*[49] β-Hydroxy-olefins fragment at about 500°C to carbonyl compounds and olefins. A cyclic mechanism seems likely. The hydrogen transfer is intramolecular; this has been shown by deuterium labelling (7.62). A particular class of β-hydroxy-olefin pyrolysis involves a retro-ene reaction in competition with the oxy-Cope rearrangement (§ 7.4).

$$\text{(structure)} \longrightarrow \text{(structure)} + Et_2C=O \qquad (7.62)$$

7. *Pyrolysis of allylic ethers.*[50]   Allylic ethers undergo the retro-ene reaction when heated at 400–600°C (7.63). The concerted nature of the fragmentation is supported by deuterium labelling experiments, and since the yields in the fragmentation can be high, the reaction has been used as a synthetic route to α-deuterio-olefins.

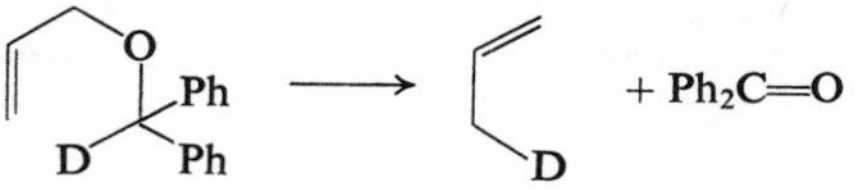

$$ \text{(structure)} \longrightarrow \text{(structure)} + Ph_2C{=}O \qquad (7.63) $$

7.9. The Cope elimination and related reactions.   There is a series of *syn* elimination reactions which probably go concertedly via a five-membered cyclic transition state (**62**, 7.64); (X = SR or NR_2; Y = O, NR or CH_2).[44] The most important of these is the Cope elimination

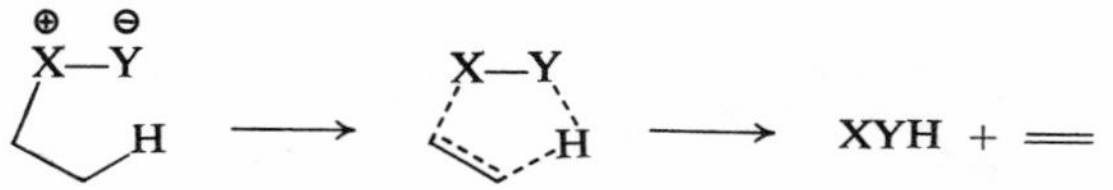

$$ \overset{\oplus\ \ominus}{X{-}Y} \ \ \longrightarrow \ \ X{-}Y \ \ \longrightarrow \ \ XYH + {=\!=} \qquad (7.64) $$

(62)

(X = NR_2, Y = O), which is a useful way of generating double bonds. Mechanistically, these eliminations bear the same relationship to the retro-ene reaction as do the ylide rearrangements and other [2, 3] sigmatropic shifts (§ 7.5) to the well established [3, 3] shifts. A simpler analogy is to compare the transition states of these reactions with those of retro-ene reactions. Both involve the movement of six electrons in an 'aromatic' transition state. In this case the electrons involved are located in the C—H σ bond, the C—X σ bond and the lone pair on Y. To draw a parallel, the retro-ene transition state is a six-membered 'benzene like' one, whereas this transition state is a five-membered 'thiophene like' one; both are aromatic. The elimination is therefore thermally allowed. On the Woodward–Hoffmann generalised picture it can be regarded as a $({}_\sigma 2_s + {}_\sigma 2_s + {}_\omega 2_s)$ reaction, as shown in fig. 7.31.

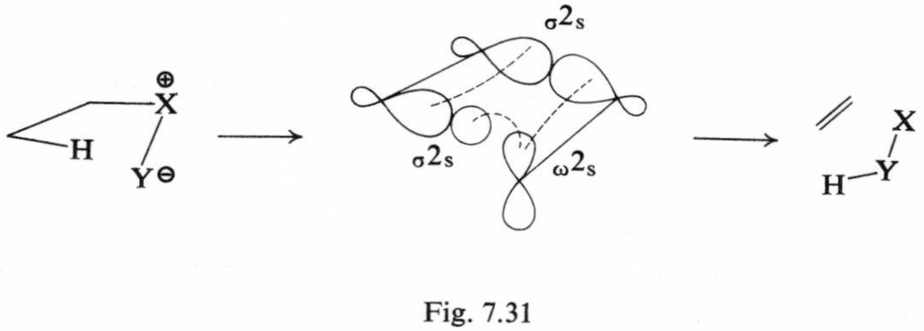

Fig. 7.31

The Cope elimination. Aliphatic tertiary amines can be converted into the corresponding amine oxides by reaction with aqueous hydrogen peroxide. The oxides fragment when heated to about 120°C and give olefins and hydroxylamines (fig. 7.32). The reaction is stereospecific in

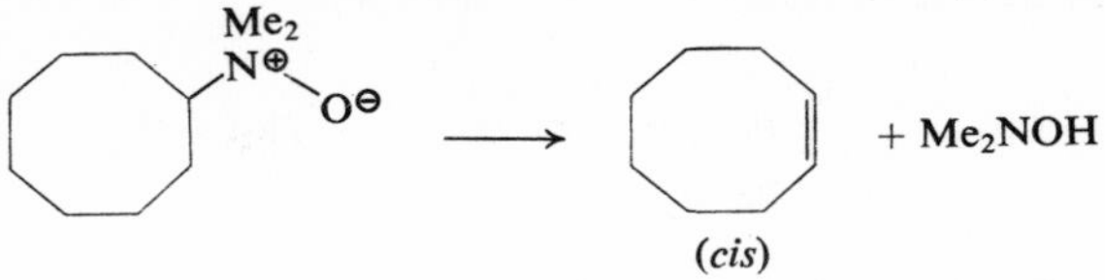

(*cis*)

Fig. 7.32

the *syn* sense, suggesting a concerted elimination; this is shown in equations 7.65 and 7.66 for the pyrolysis of the oxides (**63**) and (**64**).

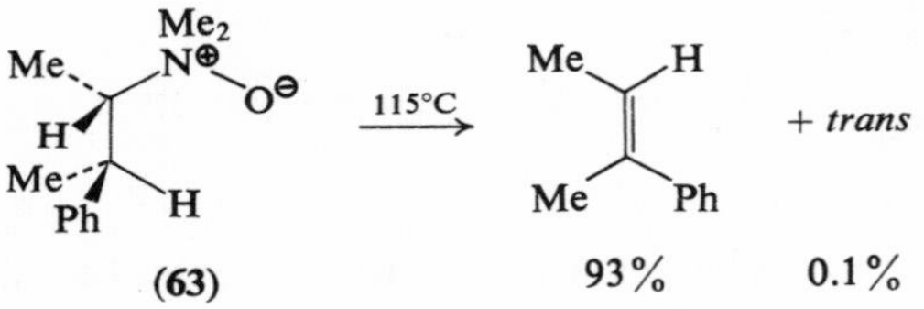

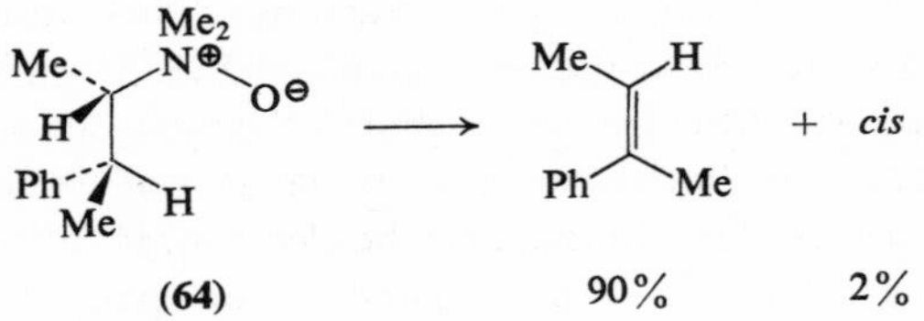

The reaction is thus very similar to the *syn* eliminations of acetates and xanthates, but goes in even milder conditions. A similar reaction is observed when the imine (**65**) is heated (7.67); the product is predominantly *cis*-cyclo-octene. A similar mechanism is probably involved.[51]

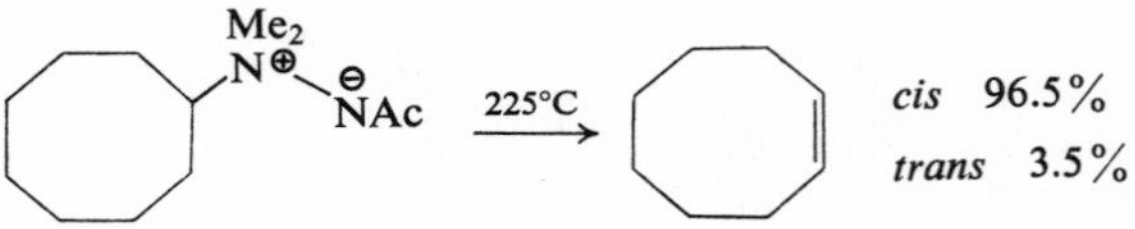

The series is completed by an ylide fragmentation which is a minor route of some Hofmann eliminations. The Hofmann elimination involves heating a quaternary ammonium hydroxide. The commonly accepted mechanism is that the hydroxide ions remove a proton β to the ammonium group (fig. 7.33). However, in special cases where the

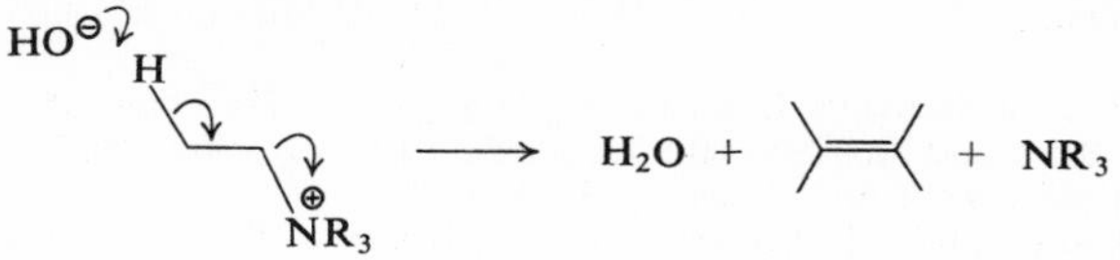

Fig. 7.33

β-hydrogen is sterically shielded from attack by an external base, an alternative ylide mechanism operates.[52] The elimination from the ylide is cyclic, like the Cope elimination, as the labelling experiment illustrated by fig. 7.34 shows: the trimethylamine produced is mono-deuterated.

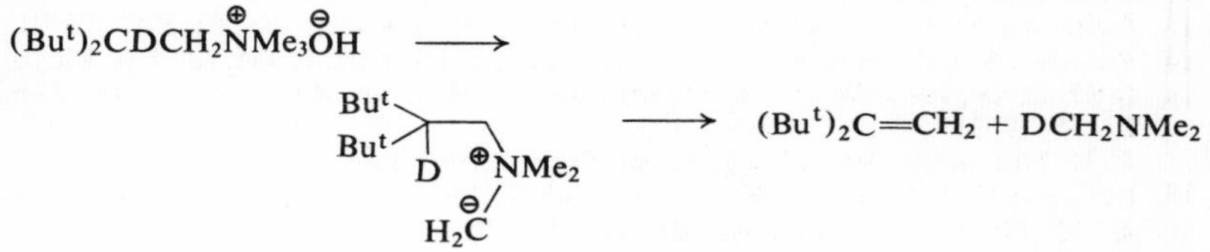

Fig. 7.34

Sulphoxide pyrolysis.[53] When simple alkyl sulphoxides are heated to about 140°C, fragmentation takes place and an olefin is formed (7.68). If the β-hydrogen is benzylic, the elimination goes at lower temperatures (80°C and above). The reaction is stereoselective and insensitive to

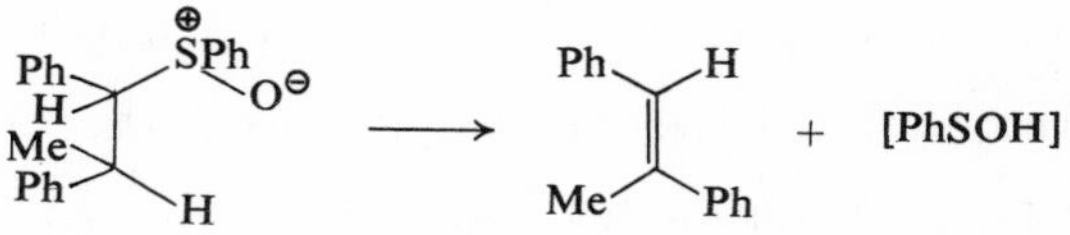

$$+ \quad [PhSOH] \qquad (7.68)$$

solvent changes at low temperatures, and is probably concerted, like the Cope elimination. At higher temperatures the stereoselectivity is lost and other products are formed; a second, stepwise mechanism increasingly competes as the temperature is raised.

REFERENCES

1. D. S. Glass, R. S. Boikess and S. Winstein, *Tetrahedron Letters*, 1966, 999, and references therein. See also ref. 43.
2. W. R. Roth, J. König and K. Stein, *Chem. Berichte*, **103**, 426 (1970).
3. D. E. McGreer and N. W. K. Chiu, *Canad. J. Chem.*, **46**, 2225 (1968).

4. W. A. Bernett, *J. Chem. Ed.*, **44**, 17 (1967); R. Hoffmann, *Tetrahedron Letters*, 1965, 3819.
5. W. Ando, *Tetrahedron Letters*, 1969, 929, and references therein. For an example of the reverse process, see J. W. Wilson and S. A. Sherrod, *Chem. Comm.*, 1968, 143.
6. J. L. M. A. Schlatmann, J. Pot and E. Havinga, *Rec. Trav. chim.*, **83**, 1173 (1964).
7. W. R. Roth and A. Friedrich, *Tetrahedron Letters*, 1969, 2607. See also J. A. Berson, *Accounts Chem. Research*, **1**, 152 (1968).
8. J. E. Baldwin and J. E. Brown, *J. Amer. Chem. Soc.*, **91**, 3647 (1969).
9. (a) L. L. Miller, R. Greisinger and R. F. Boyer, *J. Amer. Chem. Soc.*, **91**, 1578 (1969); (b) M. A. M. Boersma, J. W. DeHaan, H. Kloosterziel and L. J. M. Van de Ven, *Chem. Comm.*, 1970, 1168.
10. (a) A theoretical account of [1,2] shifts is given by N. F. Phelan, H. H. Jaffé and M. Orchin, *J. Chem. Ed.*, **44**, 626 (1967), and by H. E. Zimmerman and A. Zweig, *J. Amer. Chem. Soc.*, **83**, 1196 (1961). (b) J. J. Beggs and M. B. Meyers, *J. Chem. Soc.* (B), 1970, 930.
11. H. Hart, T. R. Rodgers and J. Griffiths, *J. Amer. Chem. Soc.*, **91**, 754 (1969).
12. H. Steinberg and F. L. J. Sixma, *Rec. Trav. chim.*, **81**, 185 (1962).
13. A. Brown, M. J. S. Dewar and W. Schoeller, *J. Amer. Chem. Soc.*, **92**, 5516 (1970).
14. Reviews: S. J. Rhoads in *Molecular Rearrangements*. part. 1, 655, ed. P. de Mayo, Interscience, 1963; W. von E. Doering and W. R. Roth, *Angew. Chem. Int. Edn*, **2**, 115 (1963).
15. R. K. Hill and N. W. Gilman, *Chem. Comm.*, 1967, 619.
16. Reviews: G. Schröder, J. F. M. Oth and R. Merényi, *Angew. Chem. Int. Edn*, **4**, 752 (1965); L. A. Paquette, *ibid.*, **10**, 11 (1971).
17. L. A. Paquette, J. R. Malpass, G. R. Krow and T. J. Barton, *J. Amer. Chem. Soc.*, **91**, 5296 (1969).
18. A. Viola, E. J. Iorio, K. K. Chen, G. M. Glover, U. Nayak and P. J. Kocienski, *J. Amer. Chem. Soc.*, **89**, 3462 (1967); see also J. A. Berson and E. J. Walsh, *ibid.*, **90**, 4729 (1968).
19. Reviews: D. S. Tarbell, *Chem. Reviews*, **27**, 495 (1940); A. Jefferson and F. Scheinmann, *Quart. Reviews*, **22**, 391 (1968); H.-J. Hansen and H. Schmid, *Chem. in Britain*, **5**, 111 (1969).
20. S. J. Rhoads and R. D. Cockroft, *J. Amer. Chem. Soc.*, **91**, 2815 (1969).
21. M. E. Synerholm, N. W. Gilman, J. W. Morgan and R. K. Hill, *J. Org. Chem.*, **33**, 1111 (1968).
22. K. C. Brannock, H. S. Pridgen and B. Thompson, *J. Org. Chem.*, **25**, 1815 (1960).
23. R. K. Hill and N. W. Gilman, *Tetrahedron Letters*, 1967, 1421, and references therein.
24. H. Kwart and J. L. Schwartz, *Chem. Comm.*, 1969, 44; H. Kwart and T. J. George, *ibid.*, 1970, 433; H. Kwart and N. Johnson, *J. Amer. Chem. Soc.*, **92**, 6064 (1970).
25. E. S. Lewis, J. T. Hill and E. R. Newman, *J. Amer. Chem. Soc.*, **90**, 662 (1968). See also D. H. R. Barton, P. D. Magnus and M. J. Pearson, *Chem. Comm.*, 1969, 550.
26. M. J. Goldstein and H. A. Judson, *J. Amer. Chem. Soc.*, **92**, 4119 (1970).
27. D. L. Garmaise, A. Uchiyama and A. F. McKay, *J. Org. Chem.*, **27**, 4509 (1962).
28. R. M. Roberts and F. A. Hussein, *J. Amer. Chem. Soc.*, **82**, 1950 (1960); J. W. Schulenberg and S. Archer, *Org. Reactions*, **14**, 1 (1965).
29. Reviews: B. Robinson, *Chem. Reviews*, **63**, 373 (1963); **69**, 227 (1969).
30. (a) J. E. Baldwin, R. E. Hackler and D. P. Kelly, *Chem. Comm.*, 1968, 538; (b) J. E. Baldwin, J. E. Brown and R. W. Cordell, *ibid.*, 1970, 31; (c) J. E. Baldwin and F. J. Urban, *ibid.*, 1970, 165; (d) J. E. Baldwin, J. DeBernardis and J. E. Patrick, *Tetrahedron Letters*, 1970, 353.

31. Review: H. E. Zimmerman in *Molecular Rearrangements*, Part 1, 345, ed. P. de Mayo, Interscience, 1963.
32. R. F. Kleinschmidt and A. C. Cope, *J. Amer. Chem. Soc.*, **66**, 1929 (1944). Review: R. A. W. Johnstone in *Mechanisms of Molecular Migrations*, vol. 2, 249, Interscience, 1969.
33. I. D. Brindle and M. S. Gibson, *Chem. Comm.*, 1969, 803.
34. A. W. Herriott and K. Mislow, *Tetrahedron Letters*, 1968, 3013; W. B. Farnham, A. W. Herriott and K. Mislow, *J. Amer. Chem. Soc.*, **91**, 6878 (1969).
35. V. Rautenstrauch, *Chem. Comm.*, 1970, 526; P. Bickart, F. W. Carson, J. Jacobus, E. G. Miller and K. Mislow, *J. Amer. Chem. Soc.*, **90**, 4869 (1968).
36. V. Rautenstrauch, *Chem. Comm.*, 1970, 4.
37. A. R. Lepley, *J. Amer. Chem. Soc.*, **91**, 1237 (1969).
38. A. R. Lepley, P. M. Cook and G. F. Willard, *J. Amer. Chem. Soc.*, **92**, 1101 (1970).
39. Reviews: H. Katz, *J. Chem. Ed.*, **48**, 84 (1971); R. C. Cookson, *Chem. in Britain*, **5**, 6 (1969). See also R. C. Cookson, J. Hudec, and M. Sharma, *Chem. Comm.*, 1971, 107.
40. F. G. Cowherd and J. L. von Rosenberg, *J. Amer. Chem. Soc.*, **91**, 2157 (1969); C. H. Campbell and M. L. H. Green, *Chem. Comm.*, 1970, 1009.
41. Review: H. M. R. Hoffmann, *Angew. Chem. Int. Edn*, **8**, 556 (1969).
42. W. Fenical, D. R. Kearns and P. Radlick, *J. Amer. Chem. Soc.*, **91**, 7771 (1969).
43. H. M. Frey and R. Walsh, *Chem. Reviews*, **69**, 103 (1969).
44. Review: C. H. DePuy and R. W. King, *Chem. Reviews*, **60**, 431 (1960).
45. P. S. Skell and W. L. Hall, *J. Amer. Chem. Soc.*, **86**, 1557 (1964).
46. C. G. Swain, R. F. W. Bader, R. M. Esteve and R. N. Griffin, *J. Amer. Chem. Soc.*, **83**, 1951 (1961); K. R. Brower, B. Gay and T. L. Konkol, *ibid.*, **88**, 1681 (1966).
47. (*a*) G. G. Smith and B. L. Yates, *J. Org. Chem.*, **30**, 2067 (1965); (*b*) B. L. Yates and J. Quijano, *ibid.*, **35**, 1239 (1970).
48. G. G. Smith and S. E. Blau, *J. Phys. Chem.*, **68**, 1231 (1964).
49. G. G. Smith and K. J. Voorhees, *J. Org. Chem.*, **35**, 2182 (1970).
50. R. C. Cookson and S. R. Wallis, *J. Chem. Soc.* (B), 1966, 1245.
51. D. G. Morris, B. W. Smith and R. J. Wood, *Chem. Comm.*, 1968, 1134.
52. J. L. Coke and M. P. Cooke, *J. Amer. Chem. Soc.*, **89**, 6701 (1967).
53. C. A. Kingsbury and D. J. Cram, *J. Amer. Chem. Soc.*, **82**, 1810 (1960); S. I. Goldberg and M. S. Sahli, *J. Org. Chem.*, **32**, 2059 (1967).

8 Substitutions, additions, and eliminations

Most organic reactions in solution are not concerted: they involve intermediates of various types. Exceptions to this broad generalisation, besides those discussed in earlier chapters, include representatives of the basic classes of acyclic organic reactions: substitutions, additions, and eliminations.

About a century ago, the first experiments were done which revealed stereochemical preferences in these classes of reactions, and by the early 1900s certain general patterns were apparent: for example, the inversion of configuration at carbon in many nucleophilic aliphatic substitutions (the Walden inversion), and the preferential *trans* arrangement of groups in addition and elimination reactions. It was these observations which provided the impetus for the development of much of modern mechanistic chemistry, in particular the mechanistic classifications of Ingold and his colleagues, and the ideas on conformational analysis which have illuminated so much of natural product chemistry.

A major advance was the linking of kinetic data with the observations on the stereochemistry of these reactions, and the recognition of two main types of process: the concerted, stereospecific reactions and the stepwise, non-stereospecific reactions. These are exemplified by the S_N2 and S_N1 mechanisms of aliphatic substitution, and by the E2 and E1 mechanisms of elimination. The modern picture differs mainly in that these are regarded as extremes of a whole spectrum of possible mechanisms in solution, the precise mechanism being susceptible to quite minor variations in the nature of the solvent, the reaction conditions, and the structures of the reactants.

8.1. An approach to the theory. The question naturally arises as to whether the theories which have so successfully predicted the stereochemical course of concerted reactions involving cyclic transition states, can also account for stereochemical preferences in reactions such as these.

254

Some reactions which show high stereoselectivity almost certainly occur stepwise; for example, the *anti*-addition of bromine to alkenes; and so orbital symmetry arguments cannot account for the overall processes. On the other hand they may explain the course of the *individual steps* of the stepwise processes (for example, the formation of a cyclic bromonium ion intermediate in the bromine addition to alkenes) and they may also be able to explain why alternative, apparently simpler, modes of reaction are not observed. To take the same example again, why is concerted *syn* addition of bromine to an alkene, via a four-membered cyclic transition state (**1**, 8.1), not the favoured mode of addition?

$$(8.1)$$

(1)

Symmetry arguments can provide an answer. Concerted *syn* addition of bromine in the manner shown is symmetry-disallowed. The transition state is of the Hückel type and involves four electrons, so is anti-aromatic; using Woodward's rule, the reaction is a $_\sigma 2_s + {}_\pi 2_s$ process, and therefore disfavoured. Cyclic transition states such as this can simply by treated by the usual theoretical approaches, and selection rules devised.

There is a greater problem with reactions such as $S_N 2$ substitution and E2 elimination, which are thought to be concerted but where the transition state cannot be 'cyclic' in the ordinary sense. It is much more difficult to devise a simple, pictorial general theory for such reactions, even though it seems certain that symmetry arguments should apply to these reactions as well. It can also be argued that solvation plays such a dominant role in ionic processes of this type that a theory which effectively ignores it is of very limited value. Nevertheless, stereochemical preferences clearly exist for such processes, and they need to be explained.[1]

Pictorially, the simplest approach to the theory of such processes is frontier orbital theory. This has been applied with fair success to explain the course of substitutions, additions and eliminations: its use to explain the inversion of configuration in $S_N 2$ substitution is illustrated in § 8.2. The value of the theory can be increased by calculations of the electron distribution in the HOMO and LVMO for individual systems.[2]

A more general theory, more difficult to justify pictorially but easier to apply to new systems, has been devised by Fukui.[3] Basically it is identical to the 'aromatic transition state' approach described in chapter 2. Fukui's calculations show that there is a striking alternation in the preferred stereochemistry for concerted reactions which is directly related to the *number of electrons* participating. This means that selection rules can be applied to predict the stereochemistry of concerted substitutions, additions, and eliminations, if it is borne in mind that effects of solvent and substituents may override the predictions in some processes. In order to do this, a formalised way of drawing the transition states is devised, and then a procedure adopted which is very similar to that described in chapter 2:

1. Draw the transition state as a series of s- and p-orbitals, or s–p-hybrids, overlapping in a stereochemically feasible way.
2. Put in + and − signs to achieve maximum overlap between adjacent orbitals.
3. Count the number of electrons participating in the transition state and note whether the terminal lobes of the acyclic system are in phase (same sign) or out of phase (opposite signs).
4. The predictions then are:

	terminal lobes	
Number of electrons	*in phase*	*out of phase*
0, 4, 8, . . ., 4n	unfavourable	favourable
2, 6, 10, . . ., (4n + 2)	favourable	unfavourable

The use of the theory is illustrated below for some familiar reactions which are commonly regarded as being concerted; here, no attempt is made to assess the evidence for their concerted nature.

8.2. Substitutions at saturated carbon

S_N2 *reactions*. In bimolecular nucleophilic substitution reactions, designated S_N2, the incoming nucleophile $Y^{\ominus}$ and the leaving group $X^{\ominus}$ are both associated with the central carbon in the transition state. There are two stereochemically distinct ways in which $Y^{\ominus}$ can attack the central carbon: on the same face as the leaving group, (2), which will lead to retention of configuration, or on the opposite face, (3), which will lead to inversion. It is found that S_N2 reactions invariably go with

(2)   (3)

inversion of configuration at carbon. The inversion is not simply due to electrostatic repulsion between the incoming and leaving groups; even when the groups are oppositely charged, inversion of configuration is still observed.

Frontier orbital theory can account for this stereochemical preference as follows: the attacking nucleophile has a high-lying occupied orbital which will interact with the σ* antibonding orbital of the C—X bond. The area of greatest electron density in the C—X σ bond is between the atoms and closer to the X atom than carbon (the bond being a polar one). Conversely the maximum amplitude of the σ* orbital is around the carbon atom. Overlap of the HOMO of Y⊖ with this σ* orbital is best achieved on the rear face of the carbon, as shown in fig. 8.1.; an approach from the front face is hampered by interaction between the lone pair on Y⊖ and lone pairs on X or the bonding pairs of the C—R bonds.[2]

Fig. 8.1. σ* Orbital of C—X.

The 'aromatic transition state' approach also accounts for inversion at carbon. The possible transition states for the substitution are drawn in a formalised way and signs put in so that as many overlapping lobes as possible are in phase. Transition state (**4**), involving retention of configuration, has the terminal lobes in phase; in transition state (**5**), involving inversion, the termini are out of phase. Four electrons participate, so according to the general rule, (**5**) is favoured.

(**4**)   (**5**)

S_E2 *reactions.* Bimolecular electrophilic substitution reactions might similarly lead either to retention of configuration or to inversion, depending on whether the transition state is of type (**4**) or type (**5**).

$$Y^{\oplus} + R_3CX \rightarrow R_3CY + X^{\oplus}$$

In these reactions only two electrons participate in the transition state, so the prediction is that transition state (**4**), involving retention of configuration at carbon, should be favoured.

Electrophilic substitution at saturated carbon is not so common as nucleophilic substitution, but the evidence on the stereochemistry of the bimolecular reaction suggests that it usually involves retention of configuration. This has been shown, for example, with optically active organometallic compounds; it is also suggested by the fact that electrophilic substitutions proceed rapidly at bridgehead carbons, whereas bimolecular nucleophilic substitutions do not.

A reaction which can be regarded as an S_E2 displacement is the insertion of singlet carbenes into C—H bonds. Singlet carbenes (**6**) are electrophilic because they have a vacant p-orbital; they are isoelectronic with carbonium ions.

(**6**)

Normally, unactivated aliphatic C—H bonds are too unreactive to participate in electrophilic displacements, but highly energetic carbenes can break the C—H bond. The reaction proceeds with retention of configuration at carbon, and appears to be concerted (8.2), in the case of insertion by singlet methylene.

$$(8.2)$$

S_N2' *reactions.* Nucleophilic substitution at an allylic carbon takes place very readily. Besides the normal reaction, rearrangement is sometimes observed, which indicates that the incoming nucleophile attacks at the γ carbon. If the reaction is bimolecular it is called the S_N2' reaction:

$$Y^{\ominus} + RCH{=}CHCH_2X \rightarrow RCHYCH{=}CH_2 + X^{\ominus}$$

Using the formalised representation, the transition state might involve a *syn* attack by $Y^{\ominus}$, (**7**), or an *anti*-attack, (**8**).

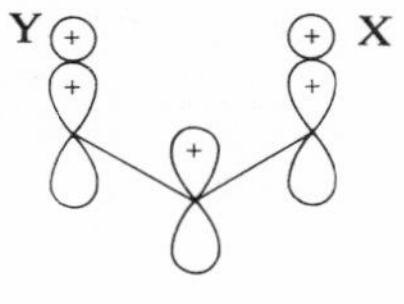

In (7) the termini are in phase, and in (8) they are out of phase. Since six electrons participate, (7) should be favoured; the attack by $Y^\ominus$ should be *syn*.

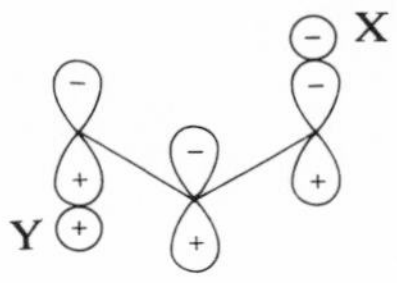

The small amount of available experimental evidence supports this prediction. For example, the allylic displacement of the benzoate group by piperidine in the cyclohexene derivative shown (8.3) is *syn*.[4]

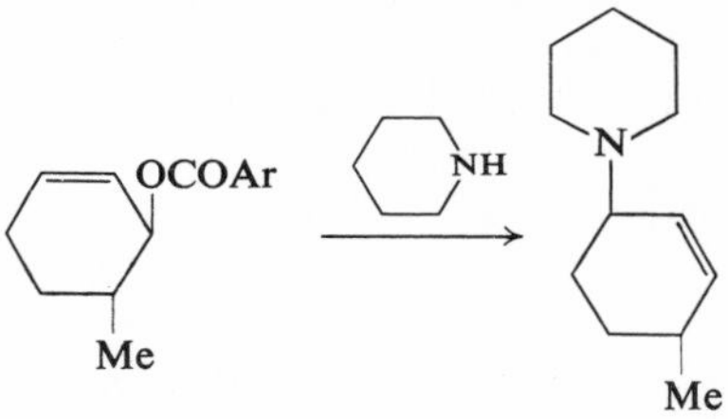

(8.3)

S_E2' *reactions*. The analogous bimolecular electrophilic substitution, with two less electrons in the transition state, should be *anti* (transition state 8). At present there is no conclusive evidence for or against this prediction.

8.3. Additions and eliminations

Concerted 1,2-*addition*. The concerted addition of a molecule XY to an olefin would involve a cyclic transition state. The rules developed in the previous chapters indicate that *syn* addition (as shown for bromine in 8.1) is not allowed. *anti*-Addition would be a $_\sigma 2_s + {}_\pi 2_a$ process, and therefore formally allowed, but geometrical constraints make the transition state difficult to attain. Such 'molecular' addition is therefore unlikely.

When X and Y are not linked in the transition state (as in ionic addition of $X^\oplus$ and $Y^\ominus$) the stereochemical possibilities are again *syn* addition (transition state 9) and *anti*-addition (transition state 10).

Four electrons participate, and the *anti*-addition (10) is therefore predicted to be favoured; *syn* addition is not favoured. Most 1,2-additions to alkenes are predominantly *anti*, although many are stepwise;

(9)   **(10)**

the completely concerted pathway represented by transition state **(10)** is a termolecular reaction, but it has been observed occasionally. Some 1,2-additions do go predominantly *syn*, for example the polar addition of hydrogen halides to alkenes, but these reactions almost certainly occur stepwise.[5]

Concerted 1,4-*addition and elimination.* 1,4-Addition to a diene might be *syn* or *anti*, and both could take place with the diene either in the *cisoid* or in the *transoid* conformation. These possibilities are represented by transition states **(11)** to **(14)**. Since six electrons participate, *syn*

$$XY + R_2C{=}CHCH{=}CR_2 \longrightarrow R_2CXCH{=}CHCR_2Y$$

(11)   **(12)**

(13)   **(14)**

addition (transition states **11** and **13**) is the favoured process. The *cisoid* diene is more likely to undergo concerted *syn* addition in practice, since here the reaction could be bimolecular, with X and Y linked in the transition state; this is stereochemically impossible for the *transoid* diene **(13)**.

Evidence on the stereochemistry of the 1,4-addition reaction is not conclusive. Although 1,4-addition to acyclic dienes is well known (8.4), it is probably a stepwise reaction in most cases, and the products are formed from addition to the *transoid* diene rather than the *cisoid*.

e.g. $Br_2 + H_2C\!=\!CHCH\!=\!CH_2$ $\longrightarrow$ $\underset{H}{\overset{BrH_2C}{\diagdown}}C\!=\!C\underset{CH_2Br}{\overset{H}{\diagup}}$ (8.4)

Cyclopentadiene, which is held in the *cisoid* conformation, adds bromine at low temperature to give mainly the *syn* adduct, *cis*-3,5-dibromocyclopentene, as the theory predicts (8.5); but there is no evidence that the reaction is concerted.[6]

$Br_2 +$ $\longrightarrow$ $Br\!\!-\!\!$ $\!\!-\!\!Br$ (8.5)

(major product)

Evidence for the reverse process, unimolecular 1,4-elimination, is more convincing. If concerted, this reaction should involve the same transition state, (**11**) or (**13**), as the addition. There are several pieces of work which show that *syn* 1,4-elimination is relatively facile and probably concerted. For example, *cis*-3,6-dimethylcyclohexadiene decomposes smoothly above 260°C to give *p*-xylene and hydrogen as the only products (8.6), whereas the *trans* dimethyl compound reacts only at appreciably higher temperatures and gives a mixture of products, mainly toluene and methane. A concerted mechanism is likely for the decomposition of the *cis* compound, but in the reaction of the *trans* isomer, radical intermediates probably participate.[7]

$\xrightarrow{260°C}$ $H_2 + Me$ Me (8.6)

Thus, the available evidence supports the view that concerted retro-additions through transition states of type (**11**) in which X and Y are linked, do take place. These are closely analogous to the retro-Diels–Alder reaction (§ 5.2).

E2 eliminations. An important group of bimolecular eliminations, E2 reactions, involves nucleophilic attack on one of the potential leaving groups, and loss of the other as an anion (8.7).

　　　　　　　　　　　　　　　　Substitutions, additions, and eliminations

$$\overset{\ominus}{N} \; \curvearrowright Y—CR_2—CR_2—\overset{\curvearrowleft}{X} \longrightarrow NY + R_2C{=}CR_2 + X^{\ominus} \qquad (8.7)$$

The preferential *anti*-periplanar arrangement of the leaving groups in E2 reactions has been regarded as one of the chief characteristics of the mechanism for a long time. For example, in E2 conditions the isomeric dibromobutanes **(15)** and **(16)** each gave mainly the olefin resulting from *anti*-elimination (8.8, 8.9).[8]

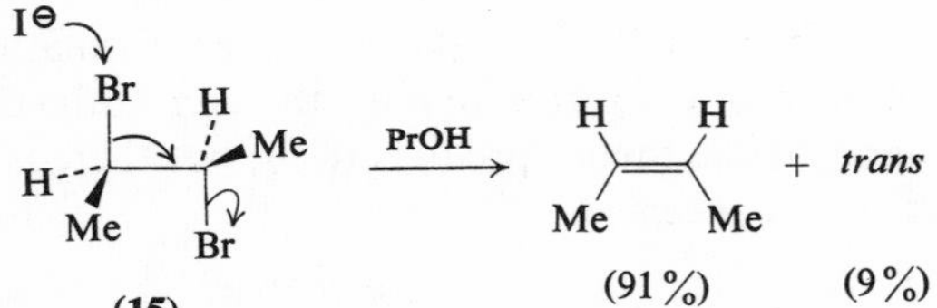

$$(8.8)$$

(15)　　　　　(91%)　　(9%)

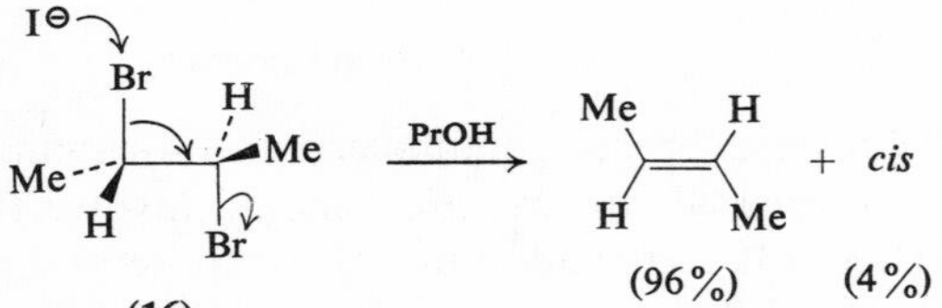

$$(8.9)$$

(16)　　　　　(96%)　　(4%)

The *anti*-periplanar principle has been particularly useful in explaining and predicting the reactions of alicyclic systems. In cyclohexane derivatives, for example, the two leaving groups must be *trans* and diaxial. Thus, the diaxial 3α,4β-dibromo-5α-cholestane **(17)** lost bromine readily when it reacted with iodide in acetone, whereas the diequatorial 3β,4α-isomer **(18)** was unreactive in the same conditions.[9].

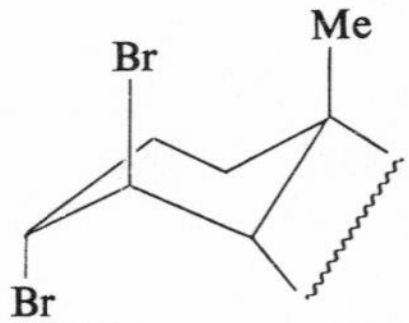

(17) 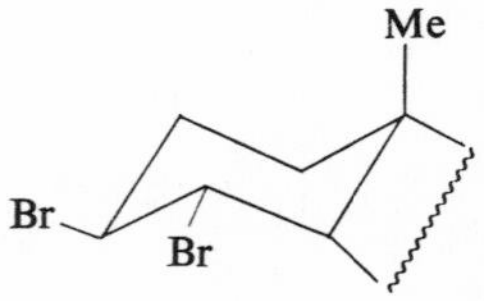　　　　　　　　　　　　　　**(18)**

The preference for *anti*-elimination can be justified on the general theory as follows. As the external nucleophile attacks Y, the bonding electron pair of the C—Y bond interacts with the electrons of the C—X bond. The interaction is formally represented for the cases where Y and

X are *syn* and *anti*, the signs being put in to indicate maximum overlap in the transition state. Since four electrons are involved, and the termini are of opposite signs, the *anti*-arrangement should be the favoured one. The process can be regarded as a type of intramolecular nucleophilic displacement.

syn   anti

It has become increasingly clear in recent years that *anti*-elimination is by no means universal, however, and there are now many examples of *syn* elimination. These examples often involve a bulky leaving group (such as trimethylammonio) and a strong, bulky base (such as t-butoxide). The energy of the transition state for *anti*-elimination may then be raised, by steric factors, to such an extent that *syn* elimination becomes preferred.[10]

Another steric factor which favours *syn* elimination may operate in certain cyclic systems when only the *cis*, and not the *trans*, substituents can achieve coplanarity (fig. 8.2).

cis groups coplanar   trans groups not coplanar

Fig. 8.2

Heterolytic fragmentations.[11] Multiple eliminations involving at least five atoms and which produce at least three separate molecules are called fragmentations. The most common type is of the general form

where *a* is an atom with a lone pair (usually nitrogen, oxygen, or carbanionic carbon), *x* is a good leaving group, and *b*, *c*, and *d* are usually, but not always, carbon.

Although there is no requirement that such processes be concerted, it is possible to write concerted mechanisms. These mechanisms can only operate if certain stereochemical features are present in the transition state. The lone pair orbital on atom a has to be suitably orientated to overlap with the developing p-orbital on b; similarly, the developing p-orbitals on atoms c and d must be parallel in the transition state. This means that the lone pair orbital on a, the b—c bond, and the d—x bond must all lie parallel. Some possible arrangements which meet these criteria are shown.

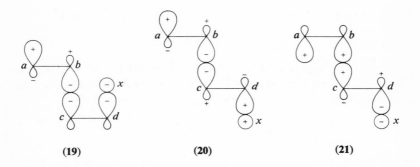

Formalised representations of the transition states (**19**) (**20**) and (**21**) (fig. 8.3) show which will be most favoured; note that transition states derived by rotation about the b—c bond, such as (**22**) from (**20**), have the same orbital representations.

Fig. 8.3

Since six electrons participate, only transition state (20) of those shown is favoured. Thus the most favourable transition states for concerted fragmentation should be those such as (20) and (22) where the lone pair on *a* and the *d—x* bond lie parallel and *anti*-periplanar.

Many fragmentable molecules do have these stereochemical features, and their fragmentations usually go much faster than those of isomers without the necessary stereochemistry. For example, the *exo*-chloride (23) reacts rapidly to give the products of fragmentation (8.10), but the *endo*-chloride (24) reacts many thousands of times more slowly and gives only substitution and elimination products.[12] The rate of ionisation of

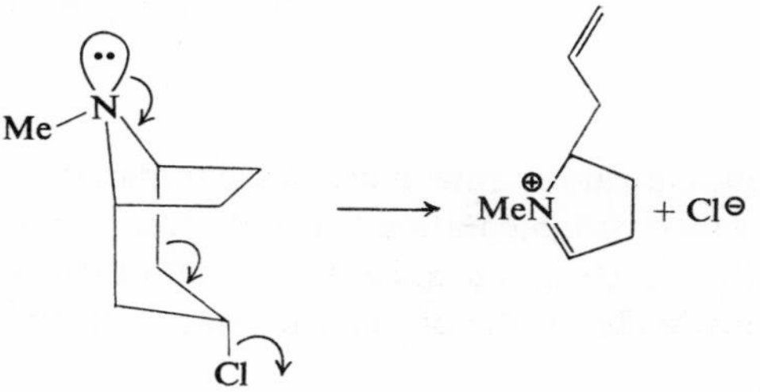

(8.10)

(23)

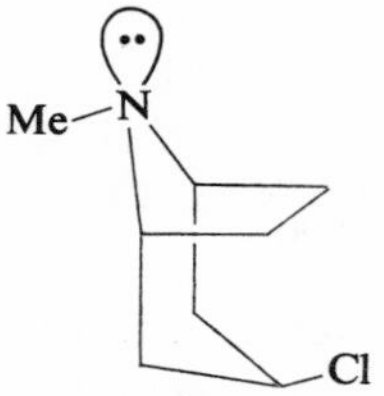

no fragmentation

(24)

the bridgehead bromide (25) is 50 000 times that of the corresponding carbon compound (26); again, concerted fragmentation is likely for (25) since it has the required structural features.[13]

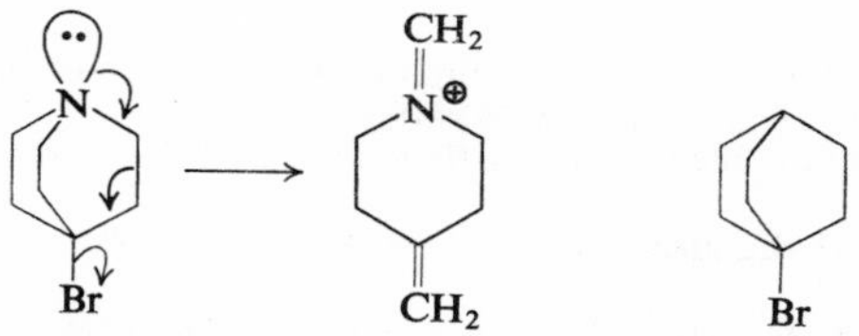

(25)   **(26)**

A special case of the reaction is the Beckmann fragmentation, which often occurs together with the Beckmann rearrangement, in derivatives of α-aminoketoximes (fig. 8.4). The reactions of α-aminoketoximes

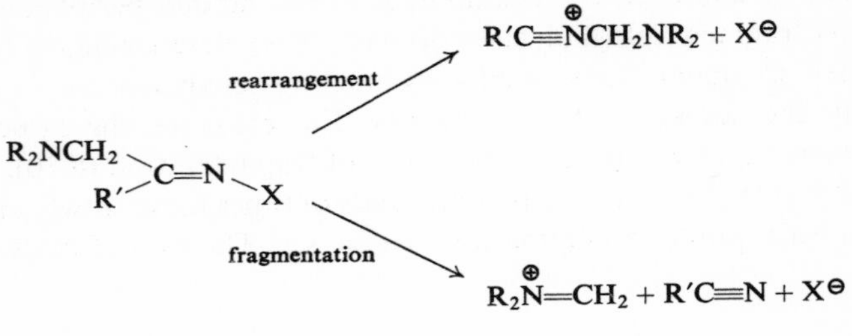

Fig. 8.4

sometimes show enormous rate increases compared with the carbon analogues in which fragmentation cannot occur. For example, the oxime acetate (**27**) reacts at a rate 10^8 times that of its carbon analogue, and gives the nitrile (**28**) as the primary product (8.11).[14]

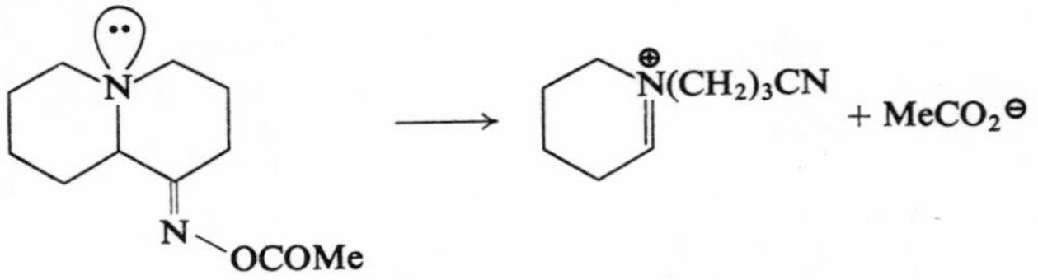

(**27**)   (**28**)

The molecular orbital predictions therefore appear to have some value and validity, even when concerted reactions choose acyclic transition states. There is scope for more theoretical and experimental work on these reactions, to determine how far the theory can usefully be applied.

REFERENCES

1. See S. I. Miller, 'Stereoselection in the Elementary Steps of Organic Reactions', in *Advances Phys. Org. Chem.*, **6**, 185 (1968).
2. L. Salem, 'Orbital Interactions in Reactions Paths', *Chem. in Britain*, **5**, 449 (1969).
3. K. Fukui and H. Fujimoto, *Bull. Chem. Soc. Japan*, **39**, 2116 (1966); **40**, 2018 (1967).
4. G. Stork and W. N. White, *J. Amer. Chem. Soc.*, **78**, 4609 (1956); G. Stork and F. H. Clarke, *ibid.*, **78**, 4619 (1956).

5. Reviews: M. J. S. Dewar and R. C. Fahey, *Angew. Chem. Int. Edn*, 3, 245 (1964); R. C. Fahey, *Topics in Stereochemistry*, 3, 237 (1968).
6. W. G. Young, H. K. Hall and S. Winstein, *J. Amer. Chem. Soc.*, 78, 4338 (1956).
7. H. M. Frey, A. Krantz and I. D. R. Stevens, *J. Chem. Soc.* (A), 1969, 1734. See also I. Fleming and E. Wildsmith, *Chem. Comm.*, 1970, 223.
8. S. Winstein, D. Pressman and W. G. Young, *J. Amer. Chem. Soc.*, 61, 1645 (1939).
9. G. H. Alt and D. H. R. Barton, *J. Chem. Soc.*, 1954, 4284.
10. D. S. Bailey and W. H. Saunders, *J. Amer. Chem. Soc.* 92, 6904 (1970).
11. Reviews: C. A. Grob and P. W. Schiess, *Angew. Chem. Int. Edn.*, 6, 1 (1967); C. A. Grob, *ibid.*, 8, 535 (1969).
12. A. T. Bottini, C. A. Grob, E. Schumacher and J. Zergenyi, *Helv. Chim. Acta*, 49, 2516 (1966).
13. P. Brenneisen, C. A. Grob, R. A. Jackson and M. Ohta, *Helv. Chim. Acta*, 48, 146 (1965).
14. C. A. Grob, H. P. Fischer, H. Link and E. Renk, *Helv. Chim. Acta*, 46, 1190 (1963).

Index